ATLA BIBLIOGRAPHY SERIES
edited by Dr. Kenneth E. Rowe

1. *A Guide to the Study of the Holiness Movement,* by Charles Edwin Jones. 1974.
2. *Thomas Merton: A Bibliography,* by Marquita E. Breit. 1974.
3. *The Sermon on the Mount: A History of Interpretation and Bibliography,* by Warren S. Kissinger. 1975.
4. *The Parables of Jesus: A History of Interpretation and Bibliography,* by Warren S. Kissinger. 1979.
5. *Homosexuality and the Judeo-Christian Tradition: An Annotated Bibliography,* by Thom Horner. 1981.
6. *A Guide to the Study of the Pentecostal Movement,* by Charles Edwin Jones. 1983.
7. *The Genesis of Modern Process Thought: A Historical Outline with Bibliography,* by George R. Lucas, Jr. 1983.
8. *A Presbyterian Bibliography,* by Harold B. Prince. 1983.
9. *Paul Tillich: A Comprehensive Bibliography ...,* by Richard C. Crossman. 1983.
10. *A Bibliography of the Samaritans,* by Alan David Crown. 1984.
11. *An Annotated and Classified Bibliography of English Literature Pertaining to the Ethiopian Orthodox Church,* by Jon Bonk. 1984.
12. *International Meditation Bibliography, 1950 to 1982,* by Howard R. Jarrell. 1984.
13. *Rabindranath Tagore: A Bibliography,* by Katherine Henn. 1985.
14. *Research in Ritual Studies: A Programmatic Essay and Bibliography,* by Ronald L. Grimes, 1985.
15. *Protestant Theological Education in America,* by Heather F. Day. 1985.
16. *Unconscious: A Guide to Sources,* by Natalino Caputi. 1985.
17. *The New Testament Apocrypha and Pseudepigrapha,* by James H. Charlesworth. 1987.
18. *Black Holiness,* by Charles Edwin Jones. 1987.
19. *A Bibliography on Ancient Ephesus,* by Richard Oster. 1987.
20. *Jerusalem, the Holy City: A Bibliography,* by James D. Purvis. 1988; Volume II, 1991.
21. *An Index to English Periodical Literature on the Old Testament and Ancient Near Eastern Studies,* by William G. Hupper. Vol. I, 1987; Vol. II, 1988; Vol. III, 1990; Vol. IV, 1990.
22. *John and Charles Wesley: A Bibliography,* by Betty M. Jarboe. 1987.
23. *A Scholar's Guide to Academic Journals in Religion,* by James Dawsey. 1988.
24. *The Oxford Movement and Its Leaders: A Bibliography of Secondary and Lesser Primary Sources,* by Lawrence N. Crumb. 1988.
25. *A Bibliography of Christian Worship,* by Bard Thompson. 1989.
26. *The Disciples and American Culture: A Bibliography of Works by Disciples of Christ Members, 1866-1984,* by Leslie R. Galbraith and Heather F. Day. 1990.
27. *The Yogacara School of Buddhism: A Bibliography,* by John Powers. 1991.
28. *The Doctrine of the Holy Spirit: A Bibliography Showing Its Chronological Development* (2 vols.), by Esther Dech Schandorff. 1992.
29. *Rediscovery of Creation: A Bibliographical Study of the Church's Response to the Environmental Crisis,* by Joseph K. Sheldon. 1992.

REDISCOVERY OF CREATION:

A Bibliographical Study of the Church's Response to the Environmental Crisis

by
Joseph K. Sheldon

ATLA Bibliography Series, No. 29

The American Theological
Library Association
and
The Scarecrow Press, Inc.
Metuchen, N.J., & London
1992

British Library Cataloguing-in-Publication data available

Library of Congress Cataloging-in-Publication data

Sheldon, Joseph Kenneth, 1943-
    Rediscovery of creation : a bibliographical study of the
Church's response to the environmental crisis / by Joseph K.
Sheldon ; foreword by Anthony Campolo.
       p. cm. -- (ATLA bibliography series ; 29)
    ISBN 0-8108-2539-2 (alk. paper)
    1.   Human ecology--Religious aspects--Christianity--
Bibliography. 2. Creation--Bibliography. 3. Nature--Religious
aspects--Christianity--Bibliography. I. Title. II. Series.
Z7799.S44 1992
[BT695.5]
016.2618'362--dc20   92-104

This book is dedicated to my parents, Donald D. and Ina M. Sheldon. Their daily walk through Creation was filled with delightful sights, sounds, smells and tastes that were savored and freely shared as they cared for their small portion of the garden.

TABLE OF CONTENTS

FOREWORD

The pendulum is about to swing. The passive quietism on social issues which has marked the last fifteen years is about to be replaced by a boisterous outburst of activism. This activism will express itself in all the cohorts of the population, but it will be most evident among those in their collegiate years. Increasingly, futurologists can see the signs of the times in a host of social indicators that range from the themes in rock music to the messages in the movies. College and university students have been the first to pick up the message, but soon it will reverberate among us all.

Get ready! The time is at hand. A social explosion is about to occur!

Whereas the '60s was a decade that had activists focused on civil rights and the '70s a time in which peace issues ruled their collective consciousness, there is little doubt that ecological concerns will be their dominant interest during the years that lie directly ahead of us. We are about to enter a period wherein preventing an all-out ecological disaster will be the major concern of the movers and shakers of society.

There is a serious question as to how the church will respond to the challenges that will be raised by an emerging Green Movement. Will the church be in the vanguard, providing a governing ideology, or will it be irrelevant to those who will embody the *geist* of the coming epoch? Will the church be riding in the locomotive pulling our society into the future or will it be in the observation car at the rear of the train, commenting on where society has been?

ix

How such questions will be answered is highly contingent upon whether or not the church is theologically prepared to respond intelligently to ecological issues. Will it have a Theology of Nature that will equip its members to enter into meaningful dialogue with those who are committed to saving Mother Earth? Will it have some answers to those who inquire as to whether God is a deity travelling with us in the Spaceship Earth or some kind of wholly transcendent being who has left behind a mechanistic universe to unwind like an abandoned clock? Will we offer a God who is involved with our plight or will we offer a God who only offers salvation in "the sweet by and by." The good news is that there are some inklings the church just might be ready for this onslaught of questions. There are some signs of hope for those of us who want a church that can answer the pressing queries posed by an anxious society with its expanding knowledge of impending ecological disaster. The good news is that a growing body of literature that speaks intelligently to issues related to environmental concerns is being amassed. There is a rediscovery of biblical themes that is providing the church with the means for developing a theology that can guide us into a moral stewardship of nature.

Much of this new thinking flows out of a doctrine of salvation that makes the human race not only the object of Christ's redemptive grace, but the instrument through which His saving work impacts the non-human spheres of creation. Giving strong emphasis to St. Paul's teaching in Romans 8:19-21, the church is coming to see the cosmic dimensions to what God has wrought through His Son, Jesus Christ, and the future that will be manifested in His Eschaton.

This book plays a major role in developing this response by the church. In it, Joseph Sheldon assembles for us a comprehensive bibliography of materials that have been developed on this theme. His book is a repository of knowledge for those who want to know just who are the major players in the theological discussion about the environment and the ways these players have contributed to this serious game in which the destiny of our planet is at stake.

Creation care is a religious calling. All of nature cries out for us to be agents of God who will participate with Him in rescuing nature from impending destruction. Acid rain, the greenhouse effect, the disruption of the ozone layer of the atmosphere, the extinction of plants with healing qualities and the wiping out of animals that have sacramental functions all cry out for the sons and daughters of God to do something. God is the ultimate voice for this imperative to be stewards of creation. And this book will help us to know where to look for the in-depth analysis of that calling and to understand how it must be lived out in everyday life.

Anthony Campolo, Ph.D.
Eastern College
St. Davids, PA

EDITOR'S FOREWORD

Since 1974 the American Theological Library Association, through its committee on publications, has undertaken responsibility for a series of published bibliographies. The series is designed to stimulate and encourage the preparation of reliable bibliographies and guides to the literature of religious studies in all of its scope and variety. Compilers are free to define their field, make their own selections, and work out internal organization as the unique demands of the subject indicate. We are pleased to publish Joseph Sheldon's *Rediscovery of Creation : A Bibliographical Study of the Church's Response to the Environmental Crisis* as number 29 in our series.

Joseph K. Sheldon, a native of Oregon, took his undergraduate degree in biology at The College of Idaho in 1966 and completed a doctorate in insect ecology at the University of Illinois in 1972. Since 1972 he has taught biology at Eastern College, St. Davids, PA. He is President of the American Entomological Society, a member of the Task Force on Global Resources of the American Scientific Affiliation, and represents Eastern College on the Academic Council and serves on the faculty of Au Sable Institute of Environmental Studies. Professor Sheldon has authored or co-authored twenty scientific articles, several of which deal with the Church's response to the environmental crisis.

Kenneth E. Rowe
Series Editor

Drew University Library
Madison, NJ 07940

PREFACE

This bibliography surveys the popular, scientific and theological literature as well as theses and dissertations that address humanity's relationship with Creation. Its focus is limited primarily to the Judeo-Christian perspective and, more particularly, to that of the Christian faith. It is limited to works published in English. Other religions associated with the Judeo-Christian tradition are also developing a theology of nature, but with the exception of a few examples to illustrate the nature of their concern, these were not included.

I obtained many of the references by searching standard theological indexing tools of the Catholic and Protestant literature such as *Religious and Theological Abstracts*, *Religion Index One*, *Religion Index Two*, and the *Catholic Periodical and Literature Index*. Other excellent sources are the bibliographies at the end of theses and dissertations, and the few published and unpublished bibliographies that are available, including Alpers 1971, Hargrove 1986, Mitcham and Grote 1984, Stange 1971, Van Dyke 1986, Innes 1987b, Engel and Bakken (forthcoming), and Wise 1989. The bibliography by Diener 1970 may also be good, but I have not been able to locate a copy for review.

I attempted to obtain copies of all cited works. When this was possible, these were deposited in the research library at Au Sable Institute of Environmental Studies, 7526 Sunset Trail N.E., Mancelona, Michigan 49659. I hope that this collection of materials will serve those of the present and future generations who are concerned with the stewardly care of Creation. It would be most helpful if those continuing to publish in this area would take it upon themselves to send copies of their works to Au Sable Institute for inclusion in the collection. Visitors are welcome to use the library.

Although this is not a fully annotated bibliography, I have quoted extensively from several of the early (prior to 1967) references to provide insight into the historical development of the theology of Creation. I have included some comments and cited many others. Because of the large number of post-1970 references, I have attempted only to categorize each work by major topic. In some cases this was difficult, and large works often dealt with several topics. In a number of cases works were listed under both the popular and theological category if it was deemed that they are important to both audiences. When I could not obtain a work through interlibrary loan for review, category assignment was based on the title and/or other authors' use of the work. In some cases these works had incomplete citations and I was forced to use the same.

A work of this type is never done in isolation, but relies on the contribuions of many others who have walked miles of dry, dusty roads before. I wish to express my gratitude to Fred Van Dyke, Dave Wise, Ron Engel and Peter Bakken for their willingness to share unpublished bibliographies, and to the library staff of Eastern College and especially Eleanor Hargreaves Mergner for her efforts in obtaining interlibrary loans. Without their help, this study would have fallen far short of its goal. Special thanks goes to Calvin B. DeWitt and Randall Frame and Eleanor Zito for reviewing the manuscript and contributing many helpful suggestions.

Joseph K. Sheldon
Eastern College
10 Fairview Drive
St. Davids, PA 19087-3696

INTRODUCTION

I began this bibliography a few years ago during a sabbatical semester from my teaching responsibilities at Eastern College. My interest in the subject springs from my training in ecology and my attempt to integrate the academic subject material of my courses and my Christian Faith.

It soon became apparent that there has been little significant dialogue among authors. Even in rather lengthy scholarly publications, the bibliographies only scratched the surface. Paul Santmire (1985b), writing from his experience with the theological literature, stated, "There is, as yet, no comprehensive review of the contemporary discussion of the theology of nature, which emerged in many countries in the late 1960's and which now includes more than twenty-five book-length studies and hundreds of articles. This is symptomatic of the discussion itself, which has generally been fragmented and occasional. Books continue to appear which betray no awareness that an international discussion of the theology of nature has been underway for twenty years" (222).

It is my hope that the material offered here will open the door for those just beginning the task of Creation care, and will as well meet the needs of those who have long tended the garden. In particular, I trust that it will encourage and facilitate the work of those in theological circles engaged in the formulation of a theology of Creation for the 21st century. It is an urgent task. Those of us working in science are keenly aware of the magnitude of the crisis. As humankind rushes ahead, bent on its quest for power and control, the groaning of Creation is increasing to an audible roar. The book of life is being snuffed out as the life support system for tens of thousands of species is expunged from the face of the earth yearly. Greenhouse effect, acid rain, the destruction of the ozone layer, and toxic waste dumps are now part of the common vocabulary, but for the most part, the Church has remained silent even though it has been suggested (White 1967, 1973) that the solution to our ecologic crisis must be religious rather than scientific or technological. If this is so, then what White and others such as Moltmann 1985, 1986, 1988 and Wright 1989 are calling for is a worldview shift within the Church.

The emerging theology of Creation is rediscovering Biblical wisdom and instruction of humankind's proper relationship with Creation. It points toward a shift from the present anthropocentric emphasis to a more Biblical theocentric view of Creation -- the picture of shalom painted in Psalm 104 and Job 38-41. By emphasizing the imminence of God in Creation as well as God's transcendent nature, humankind moves from a position of dominance to one of servanthood.

Most in the church today have yet to realize that the servant role extends to the Creation. Indeed, Kellert and Berry 1980 have found that the more frequently individuals attend religious services the less they know about the environment and the more utilitarian and dominionistic are their views. Such attitudes are not compatible with either the emerging Biblical view of Creation or a sustainable Earth.

As we share the joy of this responsibility with others, let us remember that "God gave Solomon wisdom and very great insight, and a breadth of understanding as measureless as the sand on the seashore.... He described plant life, from the cedar of Lebanon to the hyssop that grows out of walls. He also taught about animals and birds, reptiles and fish. Men of all nations came to listen to Solomon's wisdom, sent by all the kings of the world, who had heard of his wisdom" (I Kings 4:29, 33-34 [NIV]). The Church once again needs to seek this kind of wisdom and to once more speak for Creation.

CHANGING WORLD VIEWS:
THE REDISCOVERY OF CREATION

Each of us has a world view. Defined by James Sire 1976 as "a set of presuppositions (or assumptions) which we hold (consciously or subconsciously) about the basic makeup of our world" (17), one's world view determines how he or she sees the world. Walsh and Middleton 1984 and Wright 1989 extend the discussion of world views to our understanding of and proper relationship with Creation as steward.

Most assume that their world view is the "correct one." What we often fail to understand is that world views are not necessarily right or wrong -- only different. The dominant world view of contemporary western "Christian" culture differs greatly from that of the Old Testament Hebrews. That shouldn't come as much of a shock. But one result has been a profound change in the way we view our relationship with Creation. Indeed there is serious question whether the present, dominant world view is compatible with a sustainable future.

For reasons discussed by Wilkinson 1980a, there has not been an explicit theology of Creation articulated by the western Church for several hundred years. The failure of the Church to include Creation as an important part of its theology and, in contrast, to emphasize an anthropocentric theology in which nature is viewed primarily as the source of raw material to meet human needs -- something to conquer -- has led White 1967 and others to suggest that the Church and Scripture have contributed to the environmental crisis. A world view based on a theology that holds

Creation in high regard would not permit the degradation and destruction we experience today. Our world view is anthropocentric; that of the Hebrews was theocentric. We are masters of nature; the Hebrews were part of Creation. In today's western "Christian" world, nature has been desacrilized -- stripped of its holy connection with the Creator and reduced to a product of chance. Indeed, Engel 1970 argues that Scripture is not clear whether man is to exploit nature or preserve it, and Snyder 1980 states that "all the great civilized world religions [including the Judeo-Christian] remain primarily human centered" (84). In contrast, the ancient Hebrews speak of unity and shalom in Creation with man bearing the role of servant to the garden. Today's "theology of Creation" is rediscovering our rich Biblical heritage as it relates to Creation and our role as stewards.

The ability of the Church to redirect its thinking to address critical issues, including the needs of Creation, has also been called into question. Norman 1979 states:

> the Christian religion has lost the power, and the confidence, to define the areas of public debate, even in moral questions. Instead, it follows the definitions made by others. Almost no one now looks to the Church for social teaching.... The contemporary debate about world resources, over-population, pollution, or nuclear catastrophe, is according to the analyses of secular thinkers -- although the Churches tag along, offering a religious gloss to precisely the same ideas (4).

In contrast to this rather pessimistic view of the Church's ability to make a difference, one has only to scan the pages of this bibliography to realize the rapidly expanding involvement of the Church in Creation care. While many see the Church as a latecomer still struggling to climb aboard the environmental wagon, there are some segments well aboard and perhaps moving toward the driver's seat. There is no question that a leadership role is demanded of us. Indeed, Creation itself is anxiously waiting for our presence to be revealed that it might be freed from its current state of bondage and experience the glorious freedom of the Children of God (Romans 8).

THE CHURCH'S RESPONSE TO THE ENVIRONMENTAL CRISIS: AN HISTORICAL OVERVIEW

For a brief review of the church's response to the environmental crisis, see Sheldon 1989a. The general number of publications per year and significant events that may have influenced the pattern was the subject of Sheldon 1989a. A total of ten publications was found prior to 1950. The publication of White 1967 had significant influence on the nature and number of subsequent publications.

John Calvin on Genesis 2:15

> The earth was given to man, with this condition, that he should occupy himself in its cultivation.... The custody of the garden was given in charge to Adam, to show that we possess the things which God has committed to our hands, on the condition that, being content with the frugal and moderate use of them, we should take care of what shall remain. Let him who possesses a field, so partake of its yearly fruits, that he may not suffer the ground to be injured by his negligence, but let him endeavor to hand it down to posterity as he received it, or even better cultivated. Let him so feed on its fruits, that he neither dissipates it by luxury, nor permits it to be marred or ruined by neglect. Moreover, that this economy, and this diligence, with respect to those

good things which God has given us to enjoy, may flourish among us; let everyone regard himself as the steward of God in all things which he possesses. Then he will neither conduct himself dissolutely, nor corrupt by abuse those things which God requires to be preserved.

As quoted from p. 82, *Science & Christian Belief.* Vol. 2, number two. Taken from J. Calvin, *Commentary on Genesis.* First Latin edition 1554. First English Edition 1578. Present translation 1847, reprinted by Banner of Truth Publishers 1965.

1900 to 1962 -- The Years Before the Environmental Era

During the early and middle years of this century, few Christians were publishing material on Creation and humanity's relationship to it. The primary focus of the literature was on the human-God relationship and the interrelationships among humans. The following paragraphs briefly examine the contributions of some of the Christian authors who expressed concern for Creation during this period. Included are some lengthy quotes from several of the authors to illustrate the nature of their concern and the passion with which they expressed their love for Creation.

The earliest significant work I have encountered was Bailey 1915. This remarkable book, titled *The Holy Earth*, lays the groundwork for a loving and stewardly care of Creation. Although utilitarian to a degree, he clearly urges preservation. The following quotes are indicative of the content:

We did not make the earth. We have received it and its bounties. If it is beyond us, so it is divine. We have inescapable responsibilities. It is our privilege so to comprehend the use of the earth as to develop a spiritual stature. When the epoch of mere exploitation of the earth shall have worn itself out, we shall realize the heritage that remains and enter new realms of satisfaction (retrospect p. x).

We come out of the earth and we have a right to the use of the materials; and there is no danger of crass materialism if we recognize the original materials as divine and if we understand our proper relation to the creation, for then will gross selfishness in the use of them be removed. This will necessarily mean a better conception of property and of one's obligation in the use of it. We shall conceive of the earth, which is the common habitation, as inviolable. One does not act rightly toward one's fellows if one does not know how to act rightly toward the earth (2).

Bailey ends chapter 20, "The Keeping of the Beautiful Earth," with the following statement:

Merely to make the earth productive and to keep it clean and to bear a reverent regard for its products, is the special prerogative of a good agriculture and a good citizenry founded thereon; this may seem at the moment to be small and ineffective as against mad impersonal and limitless havoc, but it carries the final healing; and while the land worker will bear much of the burden on his back he will also redeem the earth (83).

Lowdermilk 1933, a forester and hydrologist, suggested that if God could have foreseen the results of centuries of abuse to Creation, there would have been an Eleventh Commandment that would have read perhaps as follows:

Thou shalt inherit the holy earth as a faithful steward, conserving its resources and productivity from generation to generation. Thou shalt safeguard thy fields from soil erosion, thy living waters from drying up, thy forests from desolation, and protect the hills from overgrazing by the herds, that thy descendants may have abundance forever. If any shall fail in this stewardship of the land, thy fruitful fields shall become sterile stony ground and wasting gullies, and thy descendants shall decrease and live in poverty or perish from off the face of the earth (as quoted from Nash 1989, pp. 97-98).

See also Helms 1984 for a discussion of Walter Lowdermilk's conversion from forester to land conservationist.

Bonhoeffer 1937 (English edition 1959) provides an insightful description of man's relationship to the Creation he is to rule. He describes it as a "freedom of dominion," stating:

this freedom of dominion directly includes our tie to the creatures who are ruled. The soil and the animals whose Lord I am are the world in which I live, without which I am not. It is my world, my earth, over which I rule. I am not free from it in the sense that my real being, my spirit requires nothing of nature, foreign to the spirit though it may be. On the contrary, in my total being, in my creatureliness, I belong to this world completely. It bears me, nourishes me, and holds me (37-38).

He goes on to state that because of the fall, man's proper relationship to Creation has been significantly changed.

We do not rule, we are ruled. The thing, the world, rules man. Man is a prisoner, a slave of the world, and his rule is illusion. Technology is the power with which the earth grips man and subdues him. And because we rule no more, we lose the ground, and then the earth is no longer our earth, and then we become strangers on the earth. We do not rule because we do not know the world as God's creation, and because we do not receive our dominion as God-given but grasp it for ourselves (38).

Raven 1940 traces the loss of the sacredness of nature to the distortion of the apostolic gospel through the first five Christian centuries. This he associates with "...the influence of Oriental and dualistic converts and with the Gnostic movements..." (63). This led to "...a denial of the worth of nature, and an insistence upon its total depravity" (64). Chapter 3, "The Renunciation of Nature," provides critical insight for those seeking to discover the true Biblical view of nature. The return of nature to its proper position within Biblical theology is of major concern to Raven. Chapter 6, "The Recovery of Nature," was written with this goal in mind. We are led through a historical tour of the Middle Ages, Renaissance and Reformation in our search. The failure of the latter to lead us back to our roots "is further proof of the extent to which the rejection of nature had been carried" (171). Indeed Raven sees the denial of nature by the church as an impediment to the spread of the gospel. He states that:

it may well be doubted whether the man who cannot find anything worshipful in the lilies or the sparrows is likely to appreciate Him who bade us consider how they grow and assured us that His Father cared for them. God is Creator as well as Redeemer; and the nature-lover is often nearer than the churchman to a true appreciation of the first article of the Creed (181).

In closing the chapter, he asserts, "the recovery of a true valuation of nature by Christendom is an event to which we can look forward with hope" (195). He then states:

God in Christ vividly apprehended strips a
man of everything: there is no room for any
word save 'God be merciful to me a sinner.'
...But it is not the whole truth: for with the gift
of Christ God does indeed give us all things --
no longer to exploit or possess but *to enjoy and
to serve* [italics mine] (196).

Dew 1950 expressed concern that "...Nature is receiving increased attention from every side of life except religion" (81). He goes on to state that:

a Christianity grounded in the Bible should
therefore develop a specifically religious attitude to Nature, distinct from that of philosophy, science or aesthetics, yet using all these
to mediate and express its fundamental faith
in the sovereignty of the Creator over the
world he has made (98).

The 1953 publication of *Nature and Man in Biblical
Thought* by E. C. Rust was the first of his several significant
works dealing with the Biblical basis of the man-nature relationship. A high view of nature is evident throughout his
works as illustrated by the following:

One aspect of man's distinctive position in
the created order is that God has given him
command over all other created things. He
gives them their names and thereby shows
that he has power to recognize their true nature and to control them. Because of this, he
is made the guardian of Paradise.
According to Genesis I, he is given dominion
over all created beings, so that he is a divine
vicegerent within the earthly realm.... He is
crowned with glory and honor and given
dominion over the works of God's hands. All
living things are put under his feet. Yet his

is a delegated authority. It is given to him by God and to God he is responsible. He and the beasts are bound together and their destinies intertwine by the intention of the creator (99-100).

Joseph Sittler has been a luminary in the developing theology of nature. His contributions began with the publication of *A Theology for Earth* in 1954 and continued for nearly 20 years. A few quotes from his early works will serve to illustrate the depth of his understanding. In *A Theology...* he identifies Creation as an area missing in contemporary theological discussions and sees neo-orthodoxy as almost repudiating the earth:

when Christian orthodoxy refuses to articulate a theology for earth, the elamant [sic] hurt of God's ancient creation is not thereby silenced. Earth's voices recollective of her lost grace and her destined redemption, will speak through one or another form of naturalism. If the Church will not have a theology for nature, then irresponsible but sensitive men will act as midwives for nature's unsilenceable meaningfulness, and enunciate a theology of nature. For earth ... unquenchably sings out her violated wholeness, and in groaning and travailing awaits with man the restoration of all things (370).

After discussing the fracture of the unity of Creation and the desacrilization of nature, Sittler 1962, in a section headed "Claiming Nature for Christ," states that:

the address of Christian thought is most weak precisely where man's ache is most strong: We have had and have, a christology of the moral soul, a christology of history, and if not a christology of the ontic, affirmations so huge as to fill the space marked out by ontological questions. But we do not have, at least not in such effective force as to have engaged the thought of the common life, a daring, penetrating, life-affirming christology of nature (183).

In 1964b Sittler states:

Man is not flung into nature as if nature were a neutral storehouse for his biological needs and intelligible operations. Man is placed in nature; and this nature is given a good and holy evaluation. And therefore man's relation to nature is not merely neutral, rational, supportive, or esthetic. It is moral -- and absolutely so. The Command of God is not that he should simply utilize the given, but that he should care for it. And the suggestion is more than faint that if he does not rightly care he shall neither rightly enjoy nor sanely use (255).

In the same year, Sittler 1964a continues:

If the creation, including our fellow creatures, is impiously used apart from a gracious primeval joy in it the very richness of the creation becomes a judgment. This has a cleansing and orderly meaning for everything in the world of nature, from the sewage we dump into our streams to the cosmic sewage we dump into the fallout.

Abuse is use without grace; it is always a failure in the counterpoint of use and enjoyment. When things are not used in ways determined by joy in the things themselves, this violated potentiality of joy (timid as all things holy, but relentless and blunt in its reprisals) withdraws and leaves us, not perhaps with immediate positive damnations but with something much worse -- the wan, ghastly, negative damnations of use without joy, stuff without grace, a busy, fabricating world with the shine gone off, personal relations for the nature of which we have invented the

eloquent term, contacts, staring without behold-
ing, even fornication without finding (97).

Asselin 1954 provides a significant theological discussion of the notion of dominion linked to the image of God
concept. The supremacy of man over nature is emphasized,
but likened to the same elevated position that God has over
nature as creator and sustainer. The following quotes illustrate the thrust of the paper: "Man enjoys a transcendence
over the rest of creation somewhat analogous to that which is
proper to God" (279); "Man is raised to an existence and
destiny sublimely like that of his creator, to rule the rest of
creation, powers of nature and all....This subservience to
man was envisaged as the primordial order of things, an
original idyllic *Shalom*, subsequently lost by the fall of
man..." (285).

The Church's failure to recognize Creation within its
theology was clearly acknowledged by Raven 1955. He suggests that this exclusion has permitted humanity to claim
Creation as its own rather than God's, and denies the creatures access to the shalom of the Kingdom. He ends his short
book with the following statement:

> In any case to recover a fuller sense of the pres-
> ence of God in His world, of the oneness of the
> three modes of His action which we know as cre-
> ation, redemption and sanctification, and so to
> face the future with faith rather than fear, hope
> not despair, and love instead of ever-stimulated
> hate would be to provide the conditions for a true
> religious revival. At present we seem to spend
> much of our energies even in the churches in try-
> ing to exclude God from His own Kingdom, to
> fence that Kingdom about so that the majority of
> His creatures are shut out from it, and to claim
> proprietary rights in its possession and adminis-
> tration for ourselves. If the study of science in
> relation to religion does nothing else, it will at
> least restore to us a truer sense of our own status.
> We shall recover the ability to say with the

Psalmist, 'O Lord, how manifold are thy works! In wisdom hast thou made them all; the earth is full of thy riches' (Ps. 104:24, A.V.); and such confession may deliver us from too exclusive a concern with ourselves and our own problems (91ff.).

Stacey 1956, near the end of a paper discussing the Christian view of nature, defines the redemption of man as the key to the release of nature from the bondage of the fall. In a discussion of the Romans 8:19-22 passage, he concludes by stating, "This brings us to the last point in the enquiry. Because the creation waits for 'the revealing of the sons of God,' the redemption of man will be the *means* of the redemption of Nature" (366).

Logan 1957 provides a useful glimpse into Genesis 1-11. His evaluation of the Genesis injunction for dominion over nature portrays a loving relationship. Although humans are creatures made in the image of God with the function of dominion, here also:

lurks man's deadly peril; his high privilege tempts him to forget that his dominion is a *delegated* dominion. He is lord of creation and ruler of nature not in his own right or to work his own will; he is charged with working God's will and is responsible to God for his stewardship. Mark well this ordained relationship, for beginning in chapter 3 of Genesis and continuing to today's newspaper you may review the sad spectacle of man's age-long effort to subdue the earth to his own ends and not to God's glory (30).

Von Rad's important theological contributions of the late 1950s and early 1960s are cited frequently by those writing on the theology of nature. In *The Problem of the Hexateuch*, a 1966 translation of an earlier German work, he points out on page 152 that in the Hebrew world view, nature was conceived of quite differently from today. Although Von Rad dealt extensively with the concept of nature,

Santmire 1985b, p. 192 considers his views to be anthropocentric. His two-volume *Old Testament Theology,* translated in 1962 and 1965, is also frequently cited.

Fagley 1960 provides a detailed early examination of Christian responsibility regarding the rapidly expanding human population. He describes the book as dealing with the "dilemmas posed by the new pressures of population and the need for a more widely held and vigorously supported Christian doctrine of responsible parenthood" (preface).

Peacocke 1961 extends the concept of stewardship to include Creation and its resources. He describes humans as God's vicegerent on earth and states that:

> the increasing contemporary dominance of man over his physical environment is within the destiny which God has willed for him. But it also reminds us sternly and unequivocally that this dominance can only be fruitful and for the good of all creation if it is exercised in conscious obedience to God's will (58).

Although individuals such as those quoted above have carried a clear message of our stewardly responsibility for Creation, western Christianity has until recently been relatively silent. It has been argued, however, that a theology of Creation has remained an important part of the eastern Christian tradition (Sweetland 1975, Birch [C.] 1975, Khalil 1978, 1987). The following statement of humanity's responsibility for the nurture and redemption of the non-human Creation by Clement, a French orthodox theologian in 1958 was quoted in Rigdon 1983 as an example of the eastern theological position regarding Creation:

> If the spiritual destiny of man is inseparable from that of humanity as a whole, it is also inseparable from that of the terrestrial cosmos. The sensible universe as a whole constitutes, in fact, a prolongation of our bodies. Or rather, what is our body, if it is not the form imprinted by

our living soul on the universal 'dust' which un-
ceasingly penetrates and traverses us? There is
no discontinuity between the flesh of the world
and human flesh; the universe participates in
human nature, as it constitutes the body of humanity...Man is the personality of the cosmos,
its conscious and personal self-expression, it is
he who gives meaning to things and who has to
transfigure them. For the universe, man is the
hope to receive grace and to be united with God;
Man is also the possibility of failure and loss for
the universe. Let us recall the fundamental text
in Romans 8:22. Subject to disorder and death by
our fall, the creation waits also for man's
becoming Son of God by grace, which would
mean liberation and glory for it also. We are
responsible for the world, to the very smallest
twigs and plants. We are the world, the 'logos'
by which the world expresses itself, by which the
world speaks to God, it depends on us whether it
blasphemes or it prays, whether it becomes an
illusion or wisdom, black magic or celebration.
Only through us can the cosmos, as the
prolongation of our bodies have access to eternity. How strange all this must sound to modern
minds! This is our evil, our sin, our freedom led
astray to vamparize [sic] nature; it is we who are
responsible for the carcasses and the twisted trees
which pollution produces, it is our refusal to love
that baffles the sad eyes of so many animals. But
every time a human being becomes aware of the
cosmic significance of the eucharist, each time a
pure being receives a humble sensation with
gratitude -- whether he eats a piece of fruit or
inhales the fragrance of the earth -- a sort of joy
of eternity reverberates in the marrow of things
(51-52).

1962 to 1966 – The Birth of Environmental Awareness

The publication of Rachel Carson's *Silent Spring* in 1962 is considered by many to mark the beginning of the environmental movement in the western world that continues to sweep forward today. For the first time, the average person began to realize humanity's impact on the environment and the magnitude of the damage that already had taken place. The initial response within the church, however, was small. Unlike the secular world, there was no significant increase in the number of publications addressing environmental issues in Christian literature (Sheldon 1989a) and few Christian authors to this day have referred to *Silent Spring*.

The number of publications per year coming from the Church on the theology of Creation or other environmental issues was 10 to 15 during this period compared to five or fewer during the 1950s. Examples of important works include the following:

Sauer 1962 addresses humanity's relationship to nature as lord and servant. He states that human dominion is derived from the Kingship Christ has chosen to share with us. Humanity's use of technical and scientific abilities "is to draw the powers of nature into his service, and in so doing into the service of God" (84). We are not, however, to be a "tyrant over nature," but:

> its lord, and as lord, should guard his kingly dignity. True, we cannot yet eliminate the universal suffering of nature, but we should not increase it, and indeed, whenever possible we should lighten it. Let us not forget that the whole creation waits for *our* perfecting. One day it will share in the liberty which the glorified children of God will possess (Romans 8:19-22). That being so, its human redeemer must not, while it waits, make its chains heavier (84).

This is a book seldom quoted, but one that should certainly be examined.

Williams 1962, following a long discussion of the wilderness theme in the Bible, closes the section with a discussion of humanity's role as Creation's steward:

> Truly the stewardship of Adam for all creatures in the Park of the Great King and the redemptive assignment laid upon Noah before the Deluge is literally in man's keeping today.... For the first time in the long history of the redemptive meaning of the wilderness, it is in our age that the forest, the jungle, the plain, and the unencumbered shore, the desert, the mountain fastness, each with its myriad denizens fashioned by the hand of the creator in their natural haunts, are becoming, surely more than he now knows, necessary for the completeness of man himself, the only creature fashioned in the image and likeness of God. Man would be less than man without his fellow creatures in all their variety of divine immediacy. Man needs now some companion in the garden bigger and freer than himself (136).

Williams continues:

> Wherever we live and work, we must have in our being or refresh within us the awareness of a real wilderness, which now we are called upon not only to contemplate periodically as did the desert Fathers, but also to conserve for ourselves and our posterity as well as in the interest of the myriad creaturely forms themselves (137).

Bring 1964 links the image of God concept with that of dominion as he describes man's relationship with the rest of Creation:

The Creator has made man the center of creation; every other created being must serve him (Gen. 1:26). The beasts are said to have been brought before him to be named (Gen. 2:19), denoting man's dominion over them. This dominion is to be exercised on God's behalf and in his name; it is not man's own dominion independent of God's Lordship. It means that man is like God and subordinate to him; although existing in his image, he cannot be independent of God; he cannot be a god beside God. He must be a servant in the divine likeness, passing on God's will to other creatures...(277).

He further states,

Since the fall had brought about a complete change in creation, involving mankind, body and soul and the whole natural world as well, the healing of this ill could only be effected through a new intervention of God, restoring and at the same time recreating nature....The work of Christ then was a restoration of this created order as well as a new creation (278).

Ditmanson 1964 seeks to restore nature within the theology of the Church. He briefly describes the loss of the sacredness of nature during the Scientific Revolution and the subsequent lack of any significant theological attempt to deal with it since.

G. W. H. Lampe read a paper at the Faith and Order Commission of the World Council of Churches in 1964 titled "The New Testament Doctrine of *Ktisis*." Humanity's relationship to nature is a central theme of the paper. Our stewardly responsibility is clearly stated:

He [humankind] can exercise the Creator's Lordship towards the rest of creation as the steward and agent of God. He is thus a kind of intermediary through whom the will of the Creator is intended to work out its purposes for the lower creatures and the inanimate world (451).

The restorative effect of the redemption of humanity on the Creation is considered in light of Romans 8:18ff. When:

> sinful man is transformed by grace into the new man in Christ, then he is set free to fulfil his proper task as God's agent and steward for God's World. The sons of God may rightly act as the masters of creation. While man is in a state of sin and corruption the creation must inevitably lack purpose and significance, for it is primarily, and increasingly, through man that God acts upon it (457).

Margaret Mead, in the introduction to White 1964, *Christians in a Technological Era*, portrays the conflict between technology and proper environmental responsibility. She states, "this book is directed toward those who take religion seriously -- those who cannot conceive of a life bearing a religious label which is not infused with that religion" (11). She goes on to state that it is:

> the responsibility of one generation to the next in ensuring that no irreversible damage is done to soil or air or water, to fish or bird or forest, and the willingness of man to accept a burden perhaps even greater than that laid on humanity when our primal parents first ate of the fruit of the tree of knowledge of good and evil -- the burden of responsibility for the survival of life on earth (13).

In another important passage she states:

it is on these questions -- the stewardship of the
earth, the cultivation of the earth as a garden,
and the command to feed and heal all men -- that
the issue is joined.... Contemporary man not
only has knowledge of good and evil; he has as
well absolute power to destroy. This man-made
power of destruction lays on man a burden he has
never before experienced -- a burden, like that of
the knowledge of good and evil, from which he
cannot escape in the foreseeable future. Given
this power, the acquisition of an understanding
of the natural laws which will enable men to
construct, protect, and maintain a warless world
is a precondition to all other benefits we may reap
from our new knowledge of nature and man
himself (15).

Moule 1964 (see Moule 1967 in bibliography) provides
a detailed discussion of environmental responsibility. He
describes humanity as God's "Vice-gerent" (from *gero* =
manage) and believes:

man is responsible before God for nature. As
long as man refuses to play the part assigned
him by God, so long the entire world of nature is
frustrated and dislocated. It is only when man is
truly fitting into his proper position as a son in
relation to God his Father that the dislocation in
the whole of nature will be reduced (10).

Whitehouse 1964 in an important work on the theology of nature states that:

the grace of God disclosed in Jesus Christ is
grace expressed dramatically and historically;
it is expressed in dealings between God
(characterized in anthropomorphic symbolism
and then acknowledged to be present among men
in a fellowman) and the human race. The phys-

ical universe is a theatre for this drama; God's creation and mankind's realm (136).

Whitehouse believes,

> Man's dominion within his realm was so devised that his treason to God brought consequences for the realm as such.... Yet the fulfillment of his original destiny is promised through the drama of redemption, and herein lies hope for the physical universe. Something is in store for it which corresponds to the redemption of man's body (136).

Wood 1964 suggests that we should look to Jesus to derive our ideal attitude toward God's Creation. He states,

> this ideal relationship is to be discovered in Jesus. He is our perfect example. If we see Him thus we see in Him a growing revelation of the Creator, and Nature becomes an ever-increasing source of the wonder so necessary to Christian worship. We shall find in our examination of this ideal that His attitude was entirely opposite to any aloofness from Nature such as has characterized even His Church at certain periods in her history, and which is so in some quarters today (44).

Baer 1966 focuses on the need for church involvement in land use issues. He states, "the church today stands at a time of decision. If she is to remain true to her prophetic heritage, she must confront the power structures of society with a fresh and cogent ethic of land usage" (1241).

Barbour 1966 argued strongly that nature is more than a mechanistic clock working independently of God, but that God does indeed act in nature:

> ...theologians have in past centuries 'had their fingers burned' by claiming divine action at

points which were later explained in scientific terms; there is an understandable desire today to avoid any possible conflict by assigning nature to the scientist and confining religion to the realm of man's inward life. But if Christianity is radically interiorized, nature is left devoid of meaning, and the stretches of cosmic history before man's appearance are unrelated to God (454).

He feels that such statements about nature, though important, occupy a "secondary place in theology" (Barbour 1968, p. 26). Later, in discussing God's power over nature, he states that:

it is in *human life*, however, that the greatest opportunities for God's influence exist today. And it is in religious experience and historical revelation, rather than in nature apart from man, that divine initiative is most clearly manifest.... Both experience and history point to a God who acts not by coercing but by evoking the response of his creatures (463).

Thus Barbour sees man as the implement chosen by God to work through his Creation. The implications of this are developed more fully by others.

Hammerton 1967 discusses the healing and unity of Creation described in the apocalyptic literature in light of our present environmental problems. It is an important pre-White contribution that explores Creation's worship and praise of the Creator. The praise, darkened by the raped Creation, still can be heard. "It cannot be right," says Hammerton, that:

whenever men avail themselves of the earth's natural resources, they should behave like vandals and leave a permanent squalor as a mark of their gratefulness. Nor can it be right that whenever these natural resources are processed in our industrial centres, noise and dirt should inevitably foul the air.... The mining of the earth does not have to be accompanied by vandalism. It is only an inordinate desire for profit that rifles the earth with no thought for its appearance afterwards (27-28).

In contrast to the high view of Creation demonstrated by the quote from Hammerton 1967, Montefiore 1965 shows a far more utilitarian view:

If bird-lovers in this country must suffer the demise of the peregrine and kite-hawk as the price of better mutton (or if Africans choose to wipe out hippopotami for a similar reason) we cannot on moral grounds object (43).

Concern for the environment is indicated, but only insofar as it influences man's well-being.

In addition to the contributions of individuals described above, in 1965 the Faith-Man-Nature Group was formed. Its birth is described by Santmire 1970c. Joranson 1977 states:

by the middle sixties some American Christians and Jews had begun to communicate their sense of religious impoverishment in the face of runaway circumstances of environmental abuse and resource use because of the intellectual weakness of their theologies and the paralyzed state of their religious ethics. One of the first initiatives to provide channels for basic scholarship and creative thought for building an explicit, sensitive Judeo-Christian approach to the environment came with the formation of the Faith-Man-Nature Group (FMN). Starting as an outgrowth of the Research Group on Theology and Conservation of the Faculty Christian

Fellowship, FMN was formed in the fall of 1965 (175).

The group held several conferences, some of which published proceedings -- Stone 1971, Stefferud 1972, Joranson and Anderson 1973, Steffenson et al. 1973, Scherer 1973.

1967 and "The Historical Roots of Our Ecologic Crisis"

In 1967 historian Lynn White alleged that the blame for the ecologic crisis can be found in Scripture, which teaches a utilitarian and dominionistic attitude toward nature rather than a stewardly one. His ideas were first presented as a lecture to the American Association for the Advancement of Science in December 1966 and subsequently published in *Science*, March 1967, pp. 1203-7. The paper has been reprinted in many other sources. White's argument was echoed by McHarg 1969, 1970; Glacken 1970a; and Toynbee 1972, 1973, who emphasized the anti-environmental lifestyle characteristic of the "Christian" West. This view has been widely accepted and is frequently encountered today. Although credit is generally given to White, Pope 1972 points out that a similar argument was raised by Charlton Ogburn, Jr., 1966:

> while the Bible enjoins men to show mercy and charity to their own kind, it has failed to inculcate a sense of affection and responsibility for the non-human world. Worse, it has implanted in the minds of men a suspicion of the body and natural processes, and a ruthless disregard for the earth and its creatures (119).

Glacken 1967 indicates an even earlier origin by crediting it to the Zen Buddhist Daisetz Suzuki 1950. In 1961 Hooykaas, in an article titled "New Responsibility in a Scientific Age," also addressed the question. After stating that "Christianity is co-responsible for the rise of modern science" and that "Christianity is co-responsible for the rise of modern technology," he asks the question, "...is Christianity also to be held responsible for the evil elements that have been a consequence of the growth of science and technology?" (82). He answers his own question with the following statements, which are then more fully developed. "The abuses of modern science and technology are not essential to them, but they are consequences of a general tendency to idolatry in all men, Christians as well as non-Christians" (82). He also states that

> the remedy, i.e. the restoration of the right attitude to science and technology may, in principle, be found in a truly Christian teaching of these disciplines, that is in a 'free science,' free from the prejudices of philosophical systems, and in a technology elaborated by men who are lords over nature in responsibility to God, and who are not the tools of their own tools (82).

Arvill 1969 (first published in 1967) also raised the question of responsibility for our present environmental conditions in a brief discussion of the "dominion" mandate. He did not, however, find sufficient cause to blame Scripture.

Direct response to White, McHarg, and others who trace the cause of the ecological crisis back to the Judeo-Christian tradition includes Tuan 1968, Bullock 1969 (who also discusses the paper by Hardin 1968), Frair 1969, Feenstra 1969, Munro 1969, Underwood 1969 (who suggests the answer is a new kind of psychedelic mystical kind of revealed religion), Bonifazi 1970, Engel 1970, Fisher 1970, Hamilton 1970, Martensen 1970a, Moncrief 1970, Schaeffer 1970, Tuan 1970, Wright 1970, Wrightsman 1970, Harder 1971, Limburg 1971, Macquarrie 1971, Oswalt 1971, Sherrell 1971, Barr 1972, Dubos 1972, 1973a,b; Gowan and Schumaker 1973, Mountcastle 1973, Young 1974, Anderson 1975, Derr 1975b, Welbourn 1975, Coleman 1976, Bennett 1977, Jaki 1978, Khalil 1978, Rendtorff 1979, Attfield 1983a, Foltz 1984, Hiers 1984, Denig 1985, Dumbrell 1985, Rossi 1985b, Santmire 1985a, Meeker 1988, Tang 1990, Timm 1990.

The consensus is that White's and McHarg's Scriptural analysis is deficient and thus their conclusion faulty. They based their case primarily on the single passage in Genesis dealing with dominion (Genesis 1:28) and failed to consider the numerous other Scriptural teachings on the concern, love, and care for Creation. Individuals wishing to examine White's argument in detail should carefully consider Derr 1975b, who cites several of White's other papers and spoke to him in person. Derr concluded:

> I am fairly certain that 'Historical Roots' has been interpreted and used in many ways contrary to his [White's] intent.... In spite of his [White's] comment about Christianity's 'burden of guilt' in the ecological crisis, he did not mean that it was the 'cause' of that crisis or of the technological society in which the troubles arose (44).

White 1967 did suggest that we should look to religion for the solution to the ecological crisis. He proposed St. Francis of Assisi as one with a proper religious emphasis on nature (see Armstrong 1973 and Allchin 1974 for discussions of St. Francis of Assisi). Deloria 1973 turned to North American Indian religions, while Means 1967 suggested that the answer lay in a world-view shift toward the more pantheistic Eastern religions. The latter position has been echoed by others, including Snyder 1969, Smith 1972, Mellert 1973, Bennet 1974, Ice 1975. Schaeffer 1970 and Wilkinson 1980a both address the Eastern philosophies and argue convincingly that this would not provide a solution. Wilkinson 1980a does point out that Eastern Orthodox Christianity has more clearly recognized and emphasized those scriptural passages dealing with man's relationship and responsibility to nature. Ehrenfeld 1978 examines humanism, the "religion of humanity," and likewise finds it presently incapable of providing the answers necessary for proper environmental care.

That White's 1967 paper had a profound and lasting impact on the Church is without question (Sheldon 1989a).

Fully half of the 80 or so papers published by Christians during 1970 on Creation-related subjects cited White. But whether the awakening of the Body of Christ to the groaning of the Creation (Romans 8:22) was spawned by the charges of White 1967, McHarg 1969, and others, or whether it naturally emerged as part of the environmental movement of the late 1960s and early 1970s (the first Earthday was April 22, 1970), is difficult or impossible to determine. What is clear is the Spirit is stirring God's people. At the very point in history when human impact on Creation is reaching critical proportions, God's still small voice has spoken to the multitudes, and a new, exciting theology of Creation is emerging.

1967 to Present -- The Church Responds to Creation's Call

It is perhaps easiest to examine the growing involvement of the Church with Creation during the years since 1967 by focusing on events and issues that have elicited significant response and/or involvement. In some cases the response has been positive (the significant work on the development of a theology of creation; recognition by Francis Schaeffer that the environmental crisis and stewardship of creation are issues that must be addressed by the Church; establishment of Au Sable Institute of environmental studies, North American Conference on Christianity and Ecology (NACCE), and North American Coalition on Religion and Ecology (NACRE); Christian contributions to the new field of environmental ethics; Christians entering the new "healing sciences" of restoration ecology and conservation biology; the beginning of some in the "pro-life" movement to recognize the sanctity of all life, not just human life; significant commitment of denominations and interdenominational groups to include stewardship of Creation as an important theological issue; the appeal of scientists to the world's spiritual leaders to join in efforts to preserve nature; and the celebration of Earthday 1990 as a Christian Holy Day by congregations across the world. Some segments within the Church, however, have either been largely silent on the issue of environmental stewardship (those publishing on "creation science") or have expressed views that can only be interpreted as having a negative influence (cornucopian views such as represented in Beisner 1990 and some with dispensational views of theology such as James Watt [see Bratton 1983 and McKeever 1990]). The environmentally oriented New Age movement and Creation Spirituality have also drawn heavy criticism from the Church. Unfortunately some (Cumbey 1983 and Hunt 1983) have not recognized the difference between orthodox Christianity's involvement with issues involving stewardship of Creation and those associated with New Age and Creation Spirituality. Each of these areas is discussed more fully below.

The Maturing Theology of Creation

Although some early papers by Christians addressing issues involving the Creation and humanity's relationship to it were anthropocentric in their approach, there have been many solid theological contributions. The years after 1970 have been particularly rich, with many authors addressing the human/Creation relationship in clear attempts to establish a new Christian view of Creation. This "new thinking," which has been variously referred to as a Theology of Nature, Theology of Ecology, or Theology of Creation, has matured. Today we are witnessing its rapid spread throughout the Church, as many serious students of scripture grapple with this critical issue.

Some important topics that have been addressed include: the stewardly intention of the original Biblical mandate in Genesis 1 and 2; the image of God concept as it relates to humanity's role in tending Creation; the result of the fall; the meaning of shalom as it pertains to Creation; the full significance of the Sabbath; the significance of the Noahic covenant with all of Creation; stewardship of the land and the result of our disobedience; the purpose and value of the non-human Creation; the role of Christ in the redemption of Creation; Christ as the Second Adam; humanity's role in the redemption of Creation; the Kingdom of God as it relates to non-human Creation; and the new heaven and new earth.

The formulation, development, and key players of the "theology of Creation" (distinct from natural theology, which will not be dealt with in this paper) can be seen in the citations listed under "Creation Care/Theology: Theological Literature" in the "Issues and Topics of Creation Care" section.

Francis Schaeffer Responds to the Environmental Crisis

The 1972 publication of *Pollution and the Death of Man: the Christian view of ecology* by Francis Schaeffer was particularly significant because of Schaeffer's high regard in the evangelical community. For many conservatives, Schaeffer's recognition of the environmental crisis as a spiritual problem that demands a strong response from the Church gave it immediate legitemacy. Schaeffer recognized the failure of the Church to address this important theological area:

> The Christian is called upon to exhibit this dominion, but exhibit it rightly: treating the thing as having value in itself, exercising dominion without being destructive. The Church should always have taught and done this, but she has generally failed to do so, and we need to confess our failure. Francis Bacon understood this, and so have other Christians at different times, but by and large we must say that for a long, long time Christian teachers including the best orthodox theologians, have shown real poverty here (72).

In Schaeffer's opinion:

> a truly biblical Christianity has a real answer to the ecological crisis. It offers a balanced and healthy attitude to nature, arising from the truth of its creation by God; it offers the hope here and now of substantial healing in nature of some of the results of the Fall, arising from the truth of redemption in Christ (81).

Au Sable Institute of Environmental Studies

Au Sable Institute was formed in 1979 to serve as a center for Christian education in the environmental field. It is affiliated with the Christian College Coalition and serves its 80 colleges. In addition to a wide selection of courses offered, Au Sable has hosted several forums focusing on various aspects of the theology of Creation. Papers presented at four of these forums have now been published in book form, including Squiers 1982, Granberg-Michaelson 1987b, DeWitt 1991a, and Prance and DeWitt 1991. In recent years seminary students have attended Au Sable for work in both environmental science and theology of Creation. See Joranson and Butigan 1985, the preface of Granberg-Michaelson 1987c, Schicker 1988a,b, Morud 1989d, Streiffert 1989a, and Scoville 1990 for additional information and background on Au Sable.

Formation of the NACCE and NACRE

The North American Conference on Christianity and Ecology first convened in August, 1987. Its founding purpose was to enable Christians from a broad spectrum of denominations to articulate a "Christian ethic of ecology." The proceedings of the first national conference have been published (Krueger 1988) and subsequent regional meetings continue. NACCE also publishes the journal *Firmament: The Quarterly of Christian Ecology*. For additional information on NACCE, see "NACCE" under Christian Organizations with a Focus on Creation.

The North American Coalition on Religion and Ecology (NACRE) was chartered in 1989 to facilitate interfaith cooperation on local and global environmental issues. Its vision is summed up in the phrase "Caring for Creation." Its stated mission (quoted from NACRE's overview piece) is as follows:

1. NACRE helps to envision a society which truly cares for the environment

(creation). NACRE encourages people everywhere to visualize a world that is both sustainable and regenerative. To accomplish this on practical and theoretical levels, NACRE calls for an ongoing ECO-3 "Trialogue," one in which religion (ecumenism), science (ecology), and society (economics) clarify and agree on global ethical principles.

2. NACRE strives to communicate its "Caring for Creation" vision by developing practical resource materials for environmental education, by organizing conferences that communicate to leaders and the public the urgency of the present environmental crisis, and by developing new uses of media for interactive ecological collaboration.

3. NACRE works to transform our culture by developing coalitions of concerned religious, scientific and environmental leaders locally, nationally and internationally who will embody this new environmentally sensitive ethic in practical models and strategies of eco-action and in their use of the regeneration process.

For additional information on NACRE and its resources, see "NACRE" under Christian Organizations with a Focus on Creation. Also see Sims 1990.

Environmental Ethics

Christianity is under attack by secular scholars. As Van Dyke 1985 points out (p. 41) in a discussion of Aldo Leopold, perhaps the most eloquent contributor of the century:

Leopold was capable of bitter sarcasm in attacking both Judeo-Christian ethics and its adherents. 'Abraham knew exactly what the land was for: it was to drip milk and honey into Abraham's mouth. At the present moment, the assurance with which we regard this assumption is inverse to the degree of our education' (Leopold 1974:240).

"In his call for a new ethic toward the biotic community, Leopold portrayed Judeo-Christian ethics as a "primitive, deficient system which could not speak to environmental dilemmas. Therefore, at least in humanity's treatment of the land, Judeo-Christian premises were to be explicitly rejected." (Van Dyke 1985:41). Unfortunately, as Van Dyke has pointed out, this assumption has trickled down into several major undergraduate textbooks in ecology. The major thrust of Van Dyke's paper can be summarized in the following quote:

> Christians have been lazy, ignorant and apathetic about environmental concerns. But only Christians possess an ethical system strong enough to bring conviction, courage, correction and direction to the environmental dilemma (45).

This same point was made earlier by Barbour 1970.

The important link between the environment and ethics has been strengthened by the publication of the journal, *Environmental Ethics*, Vol. 1, 1979. The opportunity is ripe for significant Christian involvement at the foundational level of this new discipline.

Restoration Ecology and Conservation Biology

Recently two new fields of study, restoration ecology and conservation biology, have emerged within the environmental sciences. Both involve healing and preserving Creation, and provide excellent opportunities for young Christians to serve in the "healing sciences." Neither discipline has matured sufficiently to allow comments on the Church's impact on its formation, but it is exciting to note that a number of the graduate students entering these new fields of study are coming from Christian colleges, with Au Sable Institute playing an important role in many of their educational experiences.

Theology of Creation Enters the Pro-Life Movement

Concern over human reproduction has taken two paths. Broad recognition of the human population crisis has resulted in global effort to reduce the rate of growth until ZPG (zero population growth) is achieved at a population compatible with long-term ecological sustainability. At the same time, many in the Church have adopted a strong "pro-life" position regarding abortion while avoiding the larger issue of responsible human reproduction.

At stake is much more than protecting human life. The expanding human population and its accompanying habitat destruction is taking a great toll on all of creation with the likely extinction of one-quarter of all species currently alive within the next 25 years. Susan Bratton's forthcoming book (1992) will provide an important historical and ethical examination of the problem and a strong challenge to the Church.

Clearly the answer is found in scripture -- the recognition of the value and goodness of all forms of life. Vernon Grounds 1987 -- *Modeling Consistency*, and Ron Sider 1987 --*What Does it Mean to Be Consistently Pro-Life?* have stated in papers presented at a conference on the "Sanctity of Life" at the Evangelical Round Table, Eastern College, St. Davids, PA, on 5 June 1987, that it is impossible to be fully pro-life without addressing the impact of expanding human populations not only on the quantity but also the quality of human as well as other forms of life. See also Evangelicals for Social Action and Sider 1987.

Denominational (and Interdenominational) Involvement

A brief history of the early denominational involvement in environmental issues is presented in Elsdon 1981, pages 14-16. A quote from a paper given at the 1972 UN

Conference on Human Environment in Stockholm by Dr. Elfan Rees of the World Council of Churches (The Churches in International Affairs, 1974a) includes the following statement:

> I admit, Mr. President, that the churches were slower than you were in realizing the terrible implications of this problem. You have awakened us, but in so doing you have lit a fire you cannot extinguish. We will follow you as long as you advance, we will spur if you halt, and we will take a vociferous lead if you turn back.

Although this "vociferous lead" has been slow in coming, many denominations have now taken a strong pro-Creation stand. No attempt was made to obtain a complete list of denominational involvement, position papers, internal documents, etc. However the brief list of references cited does provide an indication of the commitment of several denominations as well as interdenominational groups.
Brethren (Johnson 1978, Bhagat 1990); **Episcopal Church** (McClintock 1971, Episcopal Church 1979, Morton 1980, Episcopal Pastoral Letter 1987); **Friends** 1988; **Lutherans** (Stange 1971, Abraham 1985, Lutheran World Federation 1985, Bente 1989, Bloomquist 1989); **Mennonite** (Redekop 1986, Agricultural Concerns Committee 1989, Meyer and Meyer 1991); **United Methodist Church** 1984, Ranck 1990; **Roman Catholic** (Crotty 1971, Ward 1973, Thavis 1988, Blewett 1990); **United Presbyterian Church in the USA** 1970, 1971, 1973, 1974, 1976, 1982; The **World Council of Churches** (Birch 1971, Carothers et al. 1972, Derr 1973, Commission of the Churches on International Affairs: Reports 1970-73, Francis and Abrecht 1976, Berkhof 1977, Abrecht 1978, Gregorios 1978, Barbour 1979, Birch 1979, Dumas 1979). In 1980, a two-volume series (Shinn 1980 and Abrecht 1980) was published by the World Council of Churches. The work discusses a broad range of scientific, technological, and environmental issues. Later publications by or concerning WCC's work with environmental is-

sues include Duchrow and Liedke 1989, Niles 1989, van Hoeven 1989b, WCC 1989, Diers 1990, *International Review of Mission* 1990 Vol. 79(April):137-206 (Spirit and mission: renewing the earth), Rajotte 1990, Sider 1990a, WCC 1990.

Debate Over Origins: Are We Fighting the Wrong War?

Timm 1986 raises an important point concerning the Creation accounts of Genesis 1 and 2. By focusing their attention on the time and method of Creation, many creationists "miss the theological messages of these rich biblical creation materials" (103).

Cal DeWitt of Au Sable Institute makes the same point when he likens Creation to a gallery of beautiful paintings. While a group of art critics and paint chemists near the painting discuss and debate the details of their composition and origin, other people are busy cutting the priceless works of art into strips, removing the paint and selling the cloth for toilet paper. And the debate (and destruction) goes on.

Symptomatic of this has been the small number of papers on care of Creation in "creationist" publications. Klotz 1971a, 1972, 1984, Geisler 1989, and Ancil 1990 represent important exceptions to this generalization. In 1984, Klotz states that:

> The consistent creationist is an environmentalist because he recognizes God as his Creator and the Creator of everything. He realizes, too, that he is but a steward with stewardship responsibilities.... The creationist seeks to preserve the good world over which God has made him steward (8).

In an unfortunate response to Klotz, Robbins 1972 states that:

> Dr. Klotz, as all ecologers do, speaks of an 'ecology crisis,' of 'serious problems' in 'our deteriorating environment' caused by people

> who 'exploit the environment, waste our ... resources and cheat future generations of their legacy' (281).

He believes:

> there is no 'ecology crisis.' The entire thesis of the ecologers is the unproved asseveration that local pollution is threatening global catastrophe. Not one bit of evidence has been offered in support of (let alone proof of) such a thesis. Talk about an 'ecology crisis' is sheer irresponsibility (282).

He concludes by stating, "I find it quite incredible that Christians should be concerned about 'pollution' of the environment in 1972" (282). The response by Klotz 1972 is surprisingly kind considering the scientific ignorance shown by Robbins and his complete lack of Scriptural understanding of the rich passages pointing to our stewardly responsibilities for Creation.

<u>Leading Scientists Appeal to the Spiritual Leaders to Join in Preserving Nature</u>

Although this bibliography attests to the growing involvement of the Church in issues of Creation stewardship/theology, many believe that to date there has been mostly rhetoric, with little visible action at the "grass-roots level." Indeed the subject of Creation stewardship has not been raised from the pulpit or in Sunday school in most congregations. Church social events are also noted by the high homage they pay to the god of the disposable utensil. Most have yet to make the mental connection that Christ's parables on stewardship extend to the global resources of the planet.

Continued deterioration of global environmental conditions resulted in 1990 in an appeal by leading world scientists urging the religious community to join in the struggle of "preserving and cherishing the Earth" (Sagan 1990, Lynch 1990). According to Sagan:

> We are close to committing -- many would argue we are already committing -- what in religious language is sometimes called crimes against creation.... We know that the well-being of our planetary environment is already a source of profound concern in your councils and congregations.... Problems of such magnitude, and solutions demanding so broad a perspective, must be recognized from the outset as having a religious as well as a scientific dimension. Mindful of our common responsibility, we scientists -- many of us long engaged in combating the environmental crisis -- urgently appeal to the world religious community to commit, in word and deed, as boldly as is required to preserve the environment of the Earth (615).

In "Response of the Spiritual Leaders to the Scientist's Appeal on the Environment," more than 250 world religious leaders responded favorably to the call for a joint effort to preserve Creation. A date in 1991 was set to continue the dialogue. The event occurred in New York on June 2-3 and was attended by scientific, religious, and political leaders. A brief description of the June 2-3 meeting is reported by Christianity Today 1991 and Sider 1991. Sider lists the religious leaders in the United States in attendance. Both mainline and evangelical leaders were involved. Sider states:

> the participants endorsed proposals for new efforts in education, public policy, and local action to reverse environmental destruction. And they issued a forceful declaration calling on all people of faith to join together to save the planet and our children 'unto the

seventh generation' from ecological catastrophe (3).

Plans are now being made for a second gathering of the "Joint Appeals" for 1992.

Earthday 1990

Although hardly a whisper was heard in the Church regarding the first Earthday in 1970, there was a significant if not resounding response twenty years later. Many congregations across the country celebrated Creation's goodness, but at the same time bowed in sorrow over the pain it bears because of human sin (Romans 8) and the failure of humanity to exercise proper stewardship of the crown jewel of the Lord's Creation. Indicative of the Church's involvement in Earthday 1990 was the number of Christian journals that dedicated a 1990 (generally the Spring or April) issue to Creation stewardship. This includes *Christian Social Action*, *Christianity and Crisis*, *Daughters of Sarah*, *Epiphany*, *Impact*, *International Review of Mission*, *Religious Education*, *Sojourners*, *The Ecumenical Review*, *Tikkum: A Bimonthly Jewish Critique of Politics, Culture & Society*, and *World Christian*.

Cornucopian Views

At the same time that the Church is discovering its theological mandate for our stewardly relationship with Creation as servant/leader, some within the Church are presenting a different message. The recent popular book *Prospects for Growth* by E. Calvin Beisner (1990) bases its arguments on cornucopian economic views that minimize the existence of any significant ecological problems. Following the secular writers and economists Julian Simon and Herman Kahn 1984, he sees Earth as underpopulated and as posessing a resource base able to meet all of our future needs. Technology and market economics together are considered sufficient to meet the emergencies of both present and future environmental problems and temporary shortfalls in resource supply. Bigger is seen as better and conservation is viewed as unnecessary. Beisner's theological views regarding Creation are his own. He apparently is unaware of the discussion and hundreds of publications that have poured forth during the past quarter-century on the theology of Creation. The book also misrepresents and/or ignores much important scientific evidence, as he carefully selects only opinions that support his position. No mention is made of the loss of thousands of species to extinction as humanity expands to exercise "dominion." Indeed this utilitarian attitude toward nature is what Cobb 1970a refers to as the "dominant cultural orthodoxy" of the day (1187). Moltmann 1985 attributes the origin of this approach to Bacon: "Ever since Francis Bacon, the relationship between human beings and nature has been continually described as the relationship of a master to a slave" (137). Harvey Cox 1965 argues in *The Secular City* that the Biblical Creation account separated humanity from nature and nature from the divine, thus contributing to Western secularization and making nature "available for man's use" (23). Cox saw this as a positive contribution. The same point was made again in 1966 (Cox 1966). Clark 1968 paints an equally utilitarian view of nature when he describes the two-fold function of technology as freeing man from servitude to matter and providing the "instrument whereby the liberated spirit of man can turn again toward the material world and dominate it in a new active way" (287). A similar picture appears in the abstract of Rehwaldt 1969 -- "The account of God's creation implies that God proposes to carry out the chief intention of the created universe through man made in the image of God. Therefore, the advancing mastery of man over nature, whether on the inorganic or the organic level, is in keeping with God's plan for man."

End-Times (Dispensational) Theologies

Christians whose dispensational theology assumes the imminent second coming of Christ and the total destruction of Earth in preparation for the Kingdom of God have also been poorly represented in both the publications on

Creation-related matters and in groups actively pursuing creation/stewardship issues. This view received national attention a few years ago in the controversial statements of Interior Secretary James Watt, who justified opening 800 million acres of federal land to corporate exploitation by quoting the scriptural injunction to "occupy the land" until Jesus returns. For a response to this position, see Douglas 1981, Farrell 1981, Bratton 1983, Oates 1983, and Curry-Roper 1990. It is assumed that because of Christ's imminent return, caring for Creation is unnecessary, since there really won't be a tomorrow to worry about. This view typically argues (contrary to many important Biblical passages) for maximum utilization of resources today and evangelization as the only meaningful activity for Christian involvement. Badke 1991 (also a dispensationalist) disagrees with those who believe that caring for creation is unimportant. Although he strongly argues for obedient stewardship of creation as a high priority among believers, he sees the growing environmental crisis as part of the unmaking of creation which ultimately will be destroyed by God. One reviewer of Badke's book *Project Earth* commented that "with friends like Badke, who needs Exxon or Saddam" (Frame 1991, p. 41).

A more extreme position is taken by McKeever 1990 in his article "Father God and Mother Nature," which was written in response to Earthday 1990. In the same issue of *End-Times News Digest*, under the heading Easter/Earth Day, a section from "The Reaper" links Earth Day to a socialist-communist plot. According to "The Reaper," "Environmentalism and the concept of 'Mother Earth' are rooted in feministic witchcraft" (7). "The Reaper" goes on to say that Earth Day 1990 was chosen "astrologically" and that "Environmentalists are often like watermelons, green on the outside, but red on the inside. It's a concealed socialistic movement" (7).

In contrast to the global destruction scenarios painted by many dispensationalists, the redemption and healing of Creation is emphasized by most theologians addressing the theology of Creation. Their focus turns to such passages as Genesis 8:21, where God makes a covenant with

Creation and promises never again to destroy all living creatures; Romans 8:19-23, which describes Creation as being "liberated from its bondage to decay and brought into the glorious freedom of the children of God" (NIV); and Colossians 1:20, which indicates that through Christ's death and resurrection, the redemption of all of Creation was accomplished and peace was made by Christ's blood shed on the cross. Just as the Kingdom of God is seen as a two-fold event (imperfect in the present and awaiting Christ's return for its fullness to be realized), so Creation will be redeemed. Christ, the Second Adam, has begun the healing process. Our role as stewards is to serve Creation until His return.

<u>New Age and Creation Spirituality</u>

The popularity and spread of New Age philosophies during the 1970s and 1980s has been obvious and of major concern to the Church. At the same time, the Institute in Culture and Creation Spirituality (ICCS) and its supporters (Matthew Fox, Thomas Berry and others) have become quite visible with their strong message of environmental spirituality. This "Creation Spirituality" has diverged substantially from orthodox Christian theology, and reference to their material has been accompanied with a warning in the Christian press (Muratore 1988, Boulton 1990). In 1989 Fox was silenced for a period by the Vatican (Brow 1989). The willingness of Fox and others to incorporate New Age philosophies also calls for caution and concern. I have made no attempt to review the abundant literature on Creation Spirituality. Examples include Berry (J. F.) 1987, Berry (T.) 1980, 1985, 1987a, 1987b, 1988a, 1988b; Blewett 1987, Doyle 1983, Fox 1979, 1980, 1983a, b; Mann 1986, McAllister 1987, Dunn 1990.

It is most unfortunate that at the very time Christendom is awakening to the Spirit's call to again take up our yoke as stewards of Creation, authors such as Cumbey 1983 and Hunt 1983 (who are justifiably concerned about the New Age) seem unaware of the Biblical basis for a Theology of Creation. Indeed Cumbey has accused several Christian leaders of New Age connections because of their expressed concerns for the care of Creation. Hexham and Peowe-Hexham 1988, in an article dealing with the New Age, seem to link recent global environmental studies (*Limits to Growth* and the *Global 2000 Report to the President*) with New Age thinking and discredit both studies. Wilkinson 1987a provides an excellent response to those who arbitrarily associate Christian concern for Creation with New Age.

Creation Spirituality and New Age both have a strong environmental focus. It is tempting to wonder whether their births may be at least partly linked to the void resulting from Western Christianity's failure to include Creation within its theology. Sittler 1954 stated, "If the Church will not have a theology for nature, then irresponsible but sensitive men will act as midwives for nature's unsilenceable meaningfulness, and enunciate a theology of nature" (370). More recently Wright 1989 refers to this possibility in a discussion of the "Cyrus Principle" based on the passage in Isaiah 45, where God used the pagan King Cyrus of Persia to accomplish His will. Wright states that "it is good news for the creation and its future that many people who do not recognize themselves as stewards of God's creation nevertheless are carrying out their dominion in a thoroughly 'Christian' manner" (177).

ISSUES AND TOPICS OF CREATION CARE: REFERENCES TO THE LITERATURE

Agriculture -- *see* Land and Agriculture

Animal and Plant References in Scripture

Moldenke and Moldenke 1952, Fisher 1972, Holmgren 1972, Gammie 1978, Hareuveni 1980, Zohary 1982, Hareuveni 1984.

Biodiversity

Williams 1970, Duncan 1976, Berry (T.) 1988b, Cobb 1988, Dannacker 1988, DeWitt 1989c,d; Squiers 1989, DeWitt 1990a,d; Prance 1990, Sheldon 1990, Yancey 1991.

Creation Care/Theology: Popular and Lay Literature

<u>Thematic Issues of Journals</u>: Since 1969 when the first popular articles began to appear in the Christian press on caring for Creation, many journals have devoted entire issues or major sections to this subject. These include *Journal of the American Scientific Affiliation* 1969 Vol 21(2):34-47 (Environment), *Christian Century (The)* 1970 Vol 94(32):906-915 (Focus on the Stewardship of Earth); *Eternity* 1970 Vol 21(5):12-17 (What Are We Doing To God's Earth?); *Event* 1970 (Man and his environment); *Modern Churchman* 1970 Vol 14(1):1-118 (Nature, Man and God); *Theology Today* 1970 Vol 27(3):255-291 (Symposium on

Technology and the Environment); *Zygon* 1970 Vol 5(4): 274-369 (Conference on Ethics and Ecology of the Institute of Religion in an Age of Science); *Lutheran Quarterly* 1971 Vol 23(4):303-387 (Reflections on theological symbols, man and nature); *Religious Education* 1971 Vol 66(1):14-61 (Symposium: "Ecology and Religious Education"); *Review and Expositor* 1972 Vol 69(1):3-76 (Ecology and the Church); *Foundations* 1974 Vol 17:99-172 (Ecology and Justice); *Journal of the American Scientific Affiliation* 1974 Vol 26(1):1-21 (Population); *Journal of the American Scientific Affiliation* 1978 Vol 30(2):73-81 (Recombinant DNA Research), Vol 32 (2)70-101 (Nuclear Energy); *Dialog* 1980 Vol 19(summer):166-198 (The Land); *American Journal of Theology and Philosophy* 1983 Vol 4(1):3-48 (Technology, Culture, and Environment); *Epiphany* 1983 Vol 3(3):2-109 (Stewardship of the Earth: A Christian Response to a Spiritual Crisis); *Epiphany* 1985 Vol 6(1):3-83 (To Be Christian Is to Be Ecologist); *Word and World* 1986 Vol 6(1):3-96 (The Land); *The Christian Century* 1987 Vol 104(16):466-474 (The Oppression of Nature); *Epiphany* 1988 Vol 8(1):6-72 (Christian Ecology); *The Egg: A Journal of Eco-Justice* 1989-90 Vol 9(4)1-15 (For the Forests); *Christian Social Action* 1990 Vol 3(8):4-32,39 (Special Issue: Trashing the Earth); *Christianity and Crisis* Vol 50(7):139-158 (God's earth and our own); *Daughters of Sarah* 1990 Vol 16(My/Je):1-30 (Earthkeeping: a Christian ecofeminist vision); *Epiphany* 1990 Vol 10(Spring):7-74 (For the transfiguration of nature: papers from the symposium on orthodoxy and ecology); *Impact* 1990 Vol 47(30): Environmental Stewardship and Missions: Caring for Creation as Part of the Great Commission; *International Review of Mission* 1990 Vol 79(April):137-206 (Spirit and mission: renewing the earth); *Religious Education* 1990 Vol 85(1):4-50 (Religious education and the integrity of creation); *Sojourners* 1990 Vol 19(2):10-29 (The Cry of Creation: the Earth's Call to Restore Relationship); *The Ecumenical Review* 1990 Vol 42(April):89-174 (Come Holy Spirit -- renew the whole creation; Giver of life -- sustain your creation); *Tikkun: A Bimonthly Jewish Critique of Politics, Culture & Society* 1990 Vol 5(Mr/Ap):48-81 (Special

Feature: Earthday 1990); *World Christian* 1990 Vol 9(4):9-19
(Learning to Love The Earth).

Popular Articles and Editorials: Accompanying the global
environmental awakening of the 1960's and early 1970's
and continuing today are numerous articles and editorials
in Christian journals, chapters in books, and reprints of
sermons calling believers to ecologically responsible lifestyles as part of their responsibility. These include Wood
1964, Cox 1966, Montefiore 1966, Baer 1969, Brueggemann
1969, Cauthen 1969, Pollard 1969, Santmire 1969, Baer 1970,
Barbour 1970, Brueggemann 1970, Carrick 1970,
Christianity Today 1970a,b; Cobb 1970a, Daetz 1970a, b;
DeWolf 1970, Carrick 1970, Gowan 1970, Hefley 1970,
Kennedy 1970, Kuhn 1970, Leliaert 1970, Masse 1970,
Megivern 1970, Miller 1970, Moody Monthly 1970, Pollard
1970, Riga 1970, Robbins 1970, Santmire 1970a, c; Shepherd
1970, Spradley 1970, Sullivan 1970, Thorpe et al. 1970, United
Presbyterian Church (U.S.A.) 1970, Widman 1970,
Williams 1970, Baer 1971, Birch 1971, Blaiklock 1971, Brown
1971, Bullock 1971, *Christianity Today* 1971, Cobb 1971, Crotty
1971, DeWolf 1971a,b; Elwood 1971, Harder 1971, Hoyer 1971,
Jones 1971, Keller 1971, Krueger 1971, Lamm 1971,
Livingston 1971, Martensen 1971, McClintock 1971,
Moellering 1971, Montefiore 1971, Overman 1971,
Paternoster 1971, Pippert 1971, Reidel 1971, Stange 1971,
United Presbyterian Church (U.S.A.) 1971, Aay 1972, Bliss
1972, Campbell 1972, Crabtree 1972, 1973; Curley 1972, Davis
1972, Hahm 1972, Houston 1972, Konvitz 1972, Mayers (M.)
1972, Mayers (R. B.) 1972, Pope 1972, Richardson 1972,
Shinn 1972, Tuck 1972, Ward 1972, Armerding 1973, Basney
1973, Cauthen 1973, Crabtree 1973, Dubos 1973a, Echlin 1973,
Faramelli 1973, Gillette 1973, *Journal of the American
Scientific Affiliation* editorial 1973, Kennedy 1973,
Klostermaier 1973, Linn 1973, Schumaker (M.) 1973, United
Presbyterian Church (U.S.A.) 1973, American Baptist
Churches in the U.S.A., Board of National Ministries 1974,
Bugbee 1974, Cobb 1974a, b; Echlin 1974, Faramelli 1974,
Gilkey 1974, LaBar 1974, Morikawa 1974, Owens 1974,
Ruether 1974, Schumacher 1974, Schwarz 1974, Skoglund

1974, United Presbyterian Church in the U.S.A. 1974, Von Rohr Sauer 1974, Wicker 1974, Young 1974, Bockmuhl 1975, Boyd 1975, Derr 1975b, Klewin 1975, Lutz 1975a, b; Morikawa 1975a, Moss 1975, Navone 1975, Reid 1975, Sweetland 1975, West 1975, Wilkinson 1975, Alpers 1976, Baldwin 1976, Birch (C.) 1976a, b; Carney 1976, Dalton 1976, Donaldson 1976, Duncan 1976, Funck 1976, Hammer 1976, Heinegg 1976, Lindsell 1976, Moss 1976, Owens (O. D.) 1976, Owens (V. S.)1976, Santmire 1976, United Presbyterian Church (U.S.A.) 1976, Asian Theologians 1977, Austin 1977a,b; Baer 1977, Bryce-Smith 1977, Crabtree 1977, Cundy 1977, Dobel 1977, Donahue 1977, Dumas 1977, Gibson 1977a,b; Gish 1977, Gomes 1977, King 1977, Marietta 1977, Passmore 1977, Peters 1977, Santmire 1977, Vanstone 1977, Birch (B. C.)1978, Bube 1978a, b; Compton 1978/79, Dumas 1978, Ette and Waller 1978, Geiger 1978, Gibson 1978, Henry 1978, Mische 1978, Moss 1978, Owens 1978, Pignone 1978, Ruether 1978, Swartley 1978, Walther 1978, Zylstra 1978, Barbour 1979, Benjamin 1979, Birch (C.) 1979, Dumas 1979, Gibson 1979, Kim 1979, Miller 1979, Nelson 1979, Paradise 1979, Snell 1979, Strivers 1979, Berry 1980, Birch (C.) 1980, Gibson 1980a, b, c; Hardin 1980, Henriot 1980, Higgins 1980, Hough 1980, Leidke 1980, Morton 1980, Preston 1980, Wilkinson 1980b, Gibson 1981, Granberg-Michaelson 1981a,b,c; Hough 1981, Robb 1981, Rossi 1981a,b; Wilkinson 1981, Ferre 1982, Gibson 1982a,b,c,d; Granberg-Michaelson 1982a,b,c; Hearn 1982, Krause 1982, Moss, 1982, Philibert 1982, Schaeffer 1982, St. John 1982, United Presbyterian Church (U.S.A.) 1982, Wilkinson 1982, Aukerman 1983, Church 1983, Cooper and Kemball-Cook 1983, Ferre 1983, Garriott 1983, Gibson 1983a,b; Kime 1983; Kime, Shaw, and Garrett 1983; Muratore 1983, Oates 1983, Owens 1983b, Reinhart 1983, Schaeffer 1983, Skolimowski 1983, Snyder 1983, Steindl-Rast 1983, Tischel 1983, Todd 1983, Willis 1983, Barney 1984, Berry 1984, Cobb 1984, Douglas 1984, Granberg-Michaelson 1984a, Hultgren 1984, Klotz 1984, Pollard 1984, Sisk 1984, Stott 1984, Abraham 1985, Ballard 1985, Birch and Rasmussen 1985, Brand 1985, Bryan and Anson 1985, Cawley 1985, Crowley 1985, Denisenko 1985, Gibson 1985a,b,c; Granberg-Michaelson 1985, Institute for Ecumenical Research 1985, Jacobson 1985,

Krueger 1985a,b; Lutheran World Federation, 7th Assembly 1985; Muratore 1985a,b,c,d; Potter 1985, Reinhart 1985a,b,c; Rossi 1985a,b; Santmire 1985a, Schumacher 1985, Sorokin and Vladimir 1985a,b; Storck 1985, Tanenbaum 1985, Teitelbaum 1985, Thorkelson 1985, Tomlin 1985, Vander Zee 1985, American Baptist Churches in the U. S. A., General Board 1986a-e; Bauman 1986, Clark 1986, Daw 1986, DeWitt 1986, Falcke 1986, Gibson 1986a,b; Granberg-Michaelson 1986, Grant 1986, Heideman 1986, Hessel 1986a, Kraps 1986, Muratore 1986a,b; Murphy 1986, Olsen-Tjensvold 1986, Sorenson 1986, Timm 1986, Watson 1986, Alvarado 1986/87, Aschliman 1987, Austin 1987b, Bagby 1987, Barbieri 1987, Berry (J. F.) 1987, Berry (T.) 1987a,b; Berry (W.) 1987, Blewett 1987, Bonner 1987, Brand 1987, Brown 1987, Bryant 1987, Cable 1987, Clapp 1987, Conroy 1987a, Craine 1987, DeWitt 1987,b,c,d,e; Dickenson 1987, Dyrness 1987, Engel 1987, Episcopal Pastoral Letter 1987, Falcke 1987, Fox 1987, Fritsch 1987a,b; Froelich 1987a, Garriott 1987, Gibson 1987a,b; Gilkey 1987, Granberg-Michaelson 1987a,b,d; Gregorios 1987b, Haenke 1987, Henry 1987a, Hodel 1987, Hogan 1987, Inglis 1987, Jackson 1987c, Jaoudi 1987, Jegen 1987a,b; Krueger 1987a,b; Lepkowski and Lepkowski 1987, Lohman 1987, Lutz 1987, MacGillis 1987b, McCrary 1987, Muratore 1987, Nichols (C.) 1987, Olson 1987, Orgon 1987, Orgon and Artaza 1987, Rasmussen 1987, Ravid 1987, Reinhart 1987, Rifkin 1987, Robbins 1987, Rossi 1987, Royer 1987, Ryan 1987, Rybeck and Pasquariello 1987, Schwarz 1987, Sears 1987, Sherwood and Franklin 1987, Shoemaker 1987, Smith (M.) 1987, St. John 1987, Steele 1987, Stoner 1987, Van Elderen 1987, Wilhelm 1987, Arnold 1988/89, Berry (W.) 1988, Birch (C.) 1988, Brinkman 1988, Catholic Bishops 1988a,b; *Christianity Today* 1988, Cubbage 1988, DeWitt 1988a,b,c; Frame 1988, Francis 1988, Gibson 1988, Gibson 1988/89, Granberg-Michaelson 1988b,c; Haenke 1988, Hebblethwaite 1988, Isenhart 1988, McCarthy 1988, Monsma 1988, Green 1988, Przewozny 1988, Rajotte 1988, Rossi 1988, Samsonov 1988, Schicker 1988a,b; Sider 1988, Smith 1988, Thavis 1988, Aay 1989, Aidan 1989, Alves Da Silva 1989, Balmer 1989, Bloomquist 1989, Brubaker and Sider 1989, Cormack 1989, Cummings 1989, De Vries 1989, DeWitt

1989a,b,c,d; Echlin 1989, Elsdon 1989, FitzMaurice 1989, Freudenberger 1989b, Gibson 1989c, Govier 1989, Granberg-Michaelson 1989a, Grandy and Ragan 1989, Halver 1989, Hart 1989, Haught 1989, Jegen 1989, Kenyon 1989, Le-Chau 1989, Lill 1989, Limouris 1989, Loftus 1989, Morud 1989a,b,c,d,e; Muchina 1989, Nash 1989, Parmenov 1989, Pitchford 1989, Plantinga 1989, Polman 1989, Rands 1989, Rondeel 1989, Rossi 1989, Sheldon 1989a,b,c; Steward 1989, Stone 1989, Streiffert 1989a,b,c; Van Hoeven 1989a, Van Leeuwen 1989, Visser 1989, Williams 1989-90, Aeschliman 1990, Ancil 1990, Baker and Bradbury 1990, Barney 1990, Blewett 1990, Campbell 1990, Castro 1990, Cizik 1990, Cobourn 1990, Conroy 1990, Cooper (R.) 1990, Damaskinos 1990, DeMott 1990, DeWitt 1990 a-d; Diers 1990, Duraisingh 1990, Ehrenfeld 1990, Elshtain 1990, Frame 1990, Gambell 1990, Gibson 1990, Graham 1990, Granberg-Michaelson 1990, Green 1990, Hiebert 1990, Hovey 1990, Hulteen 1990a,b; Humphrey 1990, Jegen 1990, Kelly 1990, Knoch 1990, Lechte 1990, Lee 1990, Linn 1990, Meed 1990, Moore and Arena 1990, Pobee 1990, Prance 1990, Rajotte 1990, Ranck 1990, Rasmussen 1990, Rossi 1990, Ryan 1990, Sheldon 1990; Sider 1990a, Sims 1990, Squires 1990, Suhor 1990, Taylor 1990, Thurber 1990, Vanackere 1990, van Drimmelen 1990, Van Geest 1990, Van Gerwen 1990, Veeraraj 1990, Clobus 1991, DeWitt 1991b,c; Dyrness 1991, Gustafson 1991, Mpanya 1991, Prance 1991, Sider 1991, Testeman 1991, Yancey 1991, Moss (R.) date?

<u>Books for the General Audience</u>: Kirby 1835, Bailey 1915, Galloway 1951, Raven 1955, Murray 1956, Teilhard de Chardin 1956, 1959, 1964; Hefner 1964, Sittler 1964a, Cox 1965, Anderson 1967, Knight 1967, Arvill 1969, Kuhns 1969, Black 1970, Church of England 1970, Elder 1970, Hamilton 1970, Montefiore 1970, Nicholson 1970, Schaeffer 1970, Taylor 1970, Cauthen 1971, Heiss and McInnis 1971, Imsland 1971, Lamm 1971, Klotz 1971b, Sherrell 1971, Stone 1971, Barbour 1972, Barnette 1972a, Berry (W.) 1972, Blackburn 1972, Carothers et al. 1972, Cobb 1972, Derrick 1972, Dubos 1972, Jackson and Dubos 1972, Jensen and Tilberg 1972, Lutz and Santmire 1972, Stefferud 1972, Allen, Ulrich, and Foti 1973;

Derr 1973, Schilling 1973, Ward 1973, Passmore 1974, Spring and Spring 1974, Birch (C.) 1975, Derr 1975a, Ferre 1976, Francis and Albrecht 1976, Linzey 1976, Paternoster 1976, Bockmuhl 1977, Finnerty 1977, Hatfield 1977, Abrecht 1978, Birch (C.) 1978, Birch and Rasmussen 1978, Higgins 1978, Jaki 1978, Jegen and Manno 1978, Nibley 1978, Beach 1979, Gladwin 1979, Rifkin with Howard 1979, Barbour 1980, Dubos 1980, Houston 1980, Nelson 1980, Sheaffer and Brand 1980, Shinn 1980, Statt and Coote 1980, Wilkinson 1980a, Berry 1981, Cesaretti and Commins 1981, Elsdon 1981, Freudenberger 1981, Gray 1981, Limburg 1981, Sittler 1981, Soleri 1981, Winter 1981, Faricy 1982, Moss 1982, Shinn 1982, Snyder 1982, Taylor 1982, Attfield 1983b, Carmody 1983, Derr 1983, Horne 1983, Owens 1983a, Ruether 1983, Wolterstorff 1983, Yoder 1983, Bodner 1984, Bodo 1984, Cochrane 1984, Hart 1984, George 1984, Granberg-Michaelson 1984c, Schwartz 1984, Sharpe and Ker 1984, Walsh and Middleton 1984, Joranson and Butigan 1985, Green 1985, Greenoak 1985, Hessel 1985, Hundertmark 1985, Massey 1985, Mattson 1985, Snyder 1985, Strachan 1985, Wolters 1985, Church of England 1986, Hessel 1986b, Porter 1986, Shi 1986, Sittler 1986, Wood 1986, Austin 1987a, Birtel 1987, Granberg-Michaelson 1987c, Gregorios 1987a, Innes 1987a, McFague 1987, Palmer et al. 1987, Weber 1987; Austin 1988a, b; Granberg-Michaelson 1988a, Krueger 1988, Duchrow and Liedke 1989, Edinburgh and Geisler 1989, Hallman 1989, Mann 1989, McDaniel 1989, Niles 1989, Wright 1989, Bhagat 1990, Cooper (T.) 1990, Dowd 1990, Freudenberger 1990, McDaniel 1990, Sider 1990b, Badke 1991, Meyer and Meyer 1991, Prance and DeWitt 1991, Regenstein 1991, Slattery 1991, Thomas 1991, Bratton 1992.

Creation Care/Theology: Theological Literature

Books and Journals: Forrester-Brown 1920, Aptowitzer 1926, Temple 1934, Dodd 1946a,b; Robinson 1946, Van Til 1946, Dew 1950, Galloway 1951, Brunner 1952, Dahl 1952, McKenzie 1952, Raven 1953, Rust 1953, Asselin 1954, Conrad 1954, Gilkey 1954, Henrey 1954-55, Hicks 1954, North 1954, Sittler 1954, Anderson 1955, Kantonen 1956, Stacey 1956,

Taylor 1958, Wilder 1959, Thompson 1960, Wingren 1961, Elert 1962, Sauer 1962, Sittler 1962, Suld 1962, von Rad 1962, Williams 1962, Brattgard 1963, McFadden 1963, Schmemann 1963, Bring 1964, Ditmanson 1964, Hefner 1964, Lampe 1964, Sittler 1964a,b; Whitehouse 1964, Birch (C.) 1965, Cobb 1965, Dantine 1965, Evdokimov 1965, Gilkey 1965, Matthews 1965, Thompson 1965, von Rad 1965, Barbour 1966, Blidstein 1966, Cobb 1966, Santmire 1966, von Rad 1966, Anderson 1967, Moule 1967, Bonifazi 1967, Hammerton 1967, Rust 1967, Barbour 1968, Maloney 1968, Santmire 1968a, b; Barr 1968-69, Fretheim 1969, Hand 1969, Hefner 1969, Nelson 1969, Rehwaldt 1969, Bonifazi 1970, Derr 1970, Engel 1970, Fisher 1970, Froehlich 1970, Martensen 1970a,b; *Modern Churchman* 1970, Murray 1970, Pelcovitz 1970, Scheffczyk 1970, Sittler 1970a,b; Vaux 1970, West 1970, Alpers 1971, DeWolf 1971a,c; Fackre 1971, Gibbs 1971a,b; Hendricks 1971, Hendry 1971, Macquarrie 1971, Platman 1971, Rust 1971, Sheets 1971, Sittler 1971, Smith 1971, Stange 1971, Teske 1971, Trible 1971, Umphrey 1971, Westermann 1971, Williams 1971, Barbour 1972, Barnette 1972a, Braaten 1972, Calhoun 1972, Clampit 1972, Cobb 1972, Dannen 1972, Derrick 1972, Dubos 1972, Fritsch 1972, Habel 1972, Hurst 1972, Kaufman 1972, Kerr 1972, Maahs 1972, Moltmann 1972, Pope 1972, Ruether 1972, Rust 1972, Sittler 1972a,b; Tuck 1972, von Rad 1972, Williams 1972, Armerding 1973, Brockway 1973, Cobb 1973, Dubos 1973b, Gowan 1973, Griffin 1973b, Habig 1973, Hendry 1973, Nelson 1973, Obayashi 1973, Reumann 1973, Robinson 1973, Rudisell 1973, Santmire 1973, Schilling 1973, Bennett 1974, Braaten 1974a,b; Davies 1974, Davis 1974, Hefner 1974, King 1974, Kottukapally 1974, Meland 1974, Scott 1974, Strauss 1974, Verghese 1974, von Rohr Sauer 1974, Westermann 1974, Anderson 1975, Dumas 1975, Macquarrie 1975, Montefiore 1975, Points 1975, Ruether 1975, Santmire 1975, Shaw 1975, Sweetland 1975, Welbourn 1975, Brockway 1976, Cobb and Griffin 1976, Dalton 1976, Goodwin 1976, Stevenson 1976, Young 1976, Austin 1977b, Bennett 1977, Berkhof 1977, Bindewald 1977, Bockmuhl 1977, Brueggemann 1977, Clark 1977, Jobling 1977, Montefiore 1977, Santmire 1977, Schwarz 1977, Tilak 1977, Vanstone 1977, Alfaro 1978, Bonifazi 1978, Compton 1978/79, Elder 1978,

Gregorios 1978, Gammie 1978, Gammie et al. 1978,
Hermisson 1978, Khalil 1978, Landes 1978, Naidoff 1978,
Nibley 1978, Olsen-Tjensvold 1978, Wink 1978, Brisbin
1979, Cox 1979, Koch 1979, Moltmann 1979, Peacocke 1979,
Root 1979, Schwarz 1979b, Addinall 1980/81, Birch (C.)
1980, Brueggemann 1980, Carmody 1980, Cobb 1980, Fuerst 1980,
Hayes 1980, Helgeland 1980, Hendry 1980, Rust 1980,
Santmire 1980, Schillebeeckx 1980, Steck 1980, Wilkinson
1980a, Birch and Cobb 1981, Gustafson 1981, McIntyre 1981,
Ruether 1981, Schecter 1981, Schillebeeckx 1981, Schwarz
1981, Wilkinson 1981, Brueggemann 1982, Cobb 1982, Davies
1982, Friend 1982, Hall 1982, Hughart 1982, Manahan 1982,
Marberry 1982, McPherson 1982, United Presbyterian
Church (U.S.A.) 1982, Santmire 1982, Trickett 1982,
Westermann 1982, Anderson 1983, Axel et al. 1983, Bruce
1983, Brueggemann 1983, Carmody 1983, Christian Ecology
Group 1983, Hens-Piazza 1983, Institute for Ecumenical
Research 1983, McPherson 1983, Pang 1983, Rigdon 1983,
Ruether 1983, Schild 1983, Stewart 1983, Tokuzen 1983, Wyatt
1983, Anderson 1984, Balasuriya 1984, Bamford 1984,
Blocher 1984, Borgmann 1984, Bratton 1984, Brueggemann
1984, Dumbrell 1984, Granberg-Michaelson 1984b, Gustafson
1984, Hart 1984, Hiers 1984, Lieder 1984, Little 1984, Lonning
1984, Meyer 1984, Santmire 1984, Sharpe and Ker 1984, Soelle
1984, Wendell 1984, Westermann 1984a,b; Wingren 1984,
Wolters 1984, Allerton 1985, Birch and Rasmussen 1985,
Cauthen 1985a,b; Denig 1985, Dumbrell 1985, Ehrenfeld and
Bentley 1985, Forbes 1985, Gaston 1985, Genesis Rabbah 1985,
Gibson 1985a, Gordis 1985, Gottwald 1985, Gowan 1985, Gray
1985, Gustafson 1985, Hall 1985, Lonning 1985, McCoy 1985,
Meeks 1985, Moltmann 1985, Pobee 1985, Santmire 1985b,
Skolimowski 1985, Snyder 1985, Toulmin 1985, Willis 1985,
Bauckham 1986, Brueggemann 1986, Buitendag 1986,
Gosling 1986, Hall 1986, Herron 1986, McDaniel 1986,
McDonagh (E.) 1986, McDonagh (S.) 1986, McPherson 1986,
Moltmann 1986, Olson 1986, Redekop 1986, Richard 1986,
Santmire 1986, Sickles 1986, Timm 1986, Anderson 1987,
Austin 1987c, Birtel 1987, Brueggemann 1987, Dyrness 1987,
Evans and Cusack 1987, Freudenberger 1987c, Fritsch 1987c,
Fuggle 1987, Goossen 1987, Granberg-Michaelson 1987c,

Gregorios 1987a,b; Hart 1987, Henry 1987b, Innes 1987b, Johnston 1987, Kahl 1987, Kantzer 1987, Klotz 1987, Loader 1987, McDaniel (J.) 1987, McDaniel (T. F.) 1987, Mertens 1987, Meye 1987, Oliver 1987, Ollenburger 1987, Peacocke 1987, Peters (K. E.) 1987, Rasmussen 1987, Rimbach 1987, Ross 1987, Schuurman 1987, Stuhlmacher 1987, Thompson 1987, Walsh 1987, Wilkinson 1987b, Birch (B.C.) 1988, de la Cruz 1988, Engle 1988, Huber 1988, McDaniel 1988, Moberly 1988, Moltmann 1988, Phipps 1988, Tanner 1988, Thomas 1988, Tillich 1988, Elsdon 1989, Lilburne 1989, Limouris 1989, McFague 1989, Zizioulas 1989, Christiansen 1990, Filippi 1990, Granberg-Michaelson 1990, Landon 1990, Wright 1990, Bridger 1990, Primavesi 1990, Rogers 1990, Wade 1990, Zizioulas 1990, DeWitt 1991a,b,c; Manahan 1991, Sider 1991, Van Leeuwen 1991, Visick 1991, Wilkinson 1991, Zerbe 1991.

Theses and Dissertations: Gilkey 1954, Hicks 1954, Allison 1960, Suld 1962, McFadden 1963, Santmire 1966, Hand 1969, Nelson 1969, Calhoun 1972, Maahs 1972, Nelson 1973, Rudisell 1973, Strauss 1974, Goodwin 1976, Bindewald 1977, Clark 1977, Rowald 1977, Tilak 1977, Boettcher 1978, Olsen-Tjensvold 1978, Yoder 1978, Cox 1979, Root 1979, Schecter 1981, Hughart 1982, Manahan 1982, Marberry 1982, McPherson 1982, Trickett 1982, Stewart 1983, Wyatt 1983, Lieder 1984, Little 1984, McDonald 1984, Gaston 1985, Benjamin 1986, Blanchard 1986, Buitendag 1986, Dobson 1986, Sickles 1986, Ross 1987, Filippi 1990, Landon 1990, Scoville 1990.

Ecofeminism

Ruether 1975, Gray 1981, Ruether 1983, Bagby 1987, Nichols (M.) 1987, Rae 1987, Zelesnik 1987, Faricy 1988, Carter 1990, Dyson 1990, Filippi 1990, Finger 1990, Ingram 1990, Primavesi 1990, Raffensperger 1990, Schmitz 1990, Tatman 1990a,b.

Ecojustice

Cauthen 1973, Faramelli 1973, *American Baptist* 1974, Faramelli 1974, *Foundations* 1974, Morikawa 1974, Skoglund 1974, Morikawa 1975a, Gibson 1976, Donahue 1977, Gibson 1977a, Gomes 1977, Dumas 1978, Yoder 1978, Hough 1980, Gibson 1982a, Dorr 1984, Abraham 1985, Cauthen 1985b, Forbes 1985, Gibson 1985a, Gottwald 1985, Hessel 1985, Shinn 1985, Hessel 1986a, Arnold 1987, Engel 1987, Niles 1987, Shoemaker 1987, Duckrow and Liedke 1989, Owens 1989, Presbyterian Church (U.S.A.) 1989, Ranck 1989, Christiansen 1990, Lee 1990, Rajotte 1990, Sider 1990a,b; Somplatsky-Jarman 1990, Meyer and Meyer 1991. Also see the bibliography *Ecology, justice, and the Christian faith: a guide to the literature* by Engel and Bakken (forthcoming) for an extensive annotated bibliography on this topic.

Economics as It Relates to Creation

Faramelli 1970,1972, 1973; Daly 1973, Gibson 1975, Coleman 1976, Stivers 1976, United Presbyterian Church in the U.S.A. 1976, Daly 1977, Tiemstra 1977, Rifkin and Howard 1979, Stivers 1979, Daly 1980a,b; Moore and Jappe 1980, Tiemstra 1981, Berry 1984, McDonald 1984, Berry 1985, Gibson 1985c, Rasmussen 1985, Daw 1986, Dobson 1986, Klay 1986, Van Dyke 1986, Gibson 1987b, Goudzwaard 1987, McDaniel (J.) 1987, Overby 1987, Rybeck and Pasquariello 1987, Smith (I.) 1987, Gibson 1988, Goudzwaard 1988, McDonagh 1988, Van Dyke 1988, Daly and Cobb 1989, Hunt 1989, Robertson 1989, Beisner 1990, Bhagat 1990, Christiansen 1990, Cooper (T.) 1990, Meyer and Meyer 1991.

Education (Environmental)

DeWolf 1971a, McClintock 1971, Platman 1971, Clampit 1972, Baer 1977, Paetkau, Harder and Sawatzky 1978, Foster and Jeays 1981, Barnett and Flora 1982a,b; Mattson 1985, Dobson 1986, Barbieri 1987, Cable 1987, Dickenson 1987, Trotter 1987, Vander Zee 1987, Malone 1989,

McFague 1989, Morud 1989a, Dalton 1990, Rogers 1990, Simmons 1990, Erickson and Erickson 1991.

Energy (Non-Nuclear) and Other Natural Resources

Baer 1969, Birch 1971, Gillette 1973, Keenan 1973, Phillips 1973, Bube 1974, Pearson 1974, United Presbyterian Church in the U.S.A. 1974, Dumas 1975, Moss 1975, Bube 1976, Funck 1976, Lindsell 1976, Bube 1977, Tiemstra 1977, Abrecht 1978 -- part 2, Birch (B. C.) 1978, Birch and Rasmussen 1978, Bube 1978a, Johnson 1978, Walther 1978, Hessel 1979, Paradise 1979, Yoder 1979, Sweeting 1980, Walker 1980, Wiant 1980, Freudenberger 1981, Tiemstra 1981, Adams 1982, Chandler 1982, Compton 1982, DeWitt 1982, Ehlers 1982, Ensigns 1982, Freudenberger 1982, Granberg-Michaelson 1982d, Halteman 1982, Malloch 1982, Platt 1982, Murphy 1982, Terman (M.R.) 1982, Blair 1983, Bube 1983a,b, Allerton 1985, Cawley 1985, *American Baptist* 1986a, Batchelor 1986, Cowap 1986, Hanson 1986, Houston 1986, Riverson 1986, Rudin 1986, Steward 1986, Gonzalas 1987, Goossen 1987, Hodel 1987, Hodgson 1987, Le Roux 1987, Lofgren 1987, Schaeffer 1987, Van Eeden 1987, Bente 1989, DeWitt 1989d, Gelderloos 1989, Regan 1989, Squires 1989, Williams 1989-90, DeWitt 1990a,d; Evans 1990, Gambell 1990, Moore 1990, Sider 1990c, Smith 1990, Meyer and Meyer 1991, Sheldon 1991.

Environmental Knowledge/Attitudes and Religious Beliefs

Kellert and Berry 1980, Foster and Jeays 1981, Dobson 1986, Humphrey 1990.

Environmental Ethics

Knudsen 1962, Milgrom 1963, Kingston 1968, Barbour 1970, Church of England 1970, McCormick 1971, Shinn 1970a,b; *Zygon* 1970, Baer 1971, Cassel 1971a, b; Faramelli 1971, Livingston 1971, McCormick 1971, Scoby 1971, Barbour 1972, Barnette 1972a,b; Berry (R. J.) 1972, Calhoun 1972, Faramelli 1972, Longwood 1972, McCormack

1972, Stone 1972, Barbour 1973, Cobb 1973, Griffin 1973a, Longwood 1973, Scherer 1973, Steffenson et al. 1973, Braaten 1974a, b; *Foundations* 1974, Passmore 1974, Dumas 1975, Gibson 1975, Rasmussen 1975, Alpers 1976, Bennett 1976, Linzey 1976, Freudenberger and Hough 1977, Gustafson 1977, Harblin 1977, Birch (B. C.) 1978, Byron 1978, Gill 1978, Henry 1978, Yoder 1978, Barbour 1979, Brisbin 1979, Cobb 1979, Everett 1979, Hessel 1979, Squadrito 1979, Young 1979, Fritsch et al. 1980, Moore and Jappe 1980, Rust 1980, Simmons 1980, Wilkinson 1980a, Derr 1981, Gustafson 1981, Stein 1981, West 1981, Aay 1982, Adkinson 1982, Gelderloos 1982, Granberg-Michaelson 1982d, Keener 1982, Loy 1982, Schwarz 1982, Shinn 1982, Skillin 1982, Squires 1982, Swift 1982, Terman (C. R.) 1982, Terman (M. R.) 1982, Voth 1982, Westphal 1982, Wilkinson 1982, Yandell 1982, Attfield 1983b,c; Bratton 1983, Elliott and Gare 1983, Ferre 1983, McPherson 1983, Bratton 1984, Falcke 1984, Foltz 1984, Gustafson 1984, Link 1984, Stivers 1984, Gustafson 1985, Shinn 1985, Van Dyke 1985, Verhey 1985, Anderson 1986, Bratton 1986, Church of England 1986, Church of Scotland 1986, Cowap 1986, French 1986, Hargrove 1986; Harrison, Stivers, and Stone 1986, McDaniel 1986, Olsen-Tjensvold 1986, Rakestraw 1986, Stivers 1986, Walker 1986, Westhelle 1986, Alvarado 1986/87, Engel 1987, Innes 1987a, Kantzer 1987, Nurnberger 1987, Schuurman 1987, Thomas 1987, Van Eeden 1987, Weber 1987, Bratton 1988, Engel (J.R.) 1988, Harakas 1988, Kay 1988, McDaniel 1988, Nash 1989, Bratton 1990, Callicott 1990, Cupitt 1990, Dunn 1990, Engel and Engel 1990, Freudenberger 1990, Gardiner 1990, Hay 1990, John Paul 1990, McKibben 1990, Sherrard 1990, Meyer and Meyer 1991, Bratton 1992.

Genetic Engineering (Biotechnology)

Ramm 1971, Herrmann 1976, Albert 1978, Bube 1978b, Erickson 1978, Geels 1978, Haas 1978, Herrmann 1978, Jappe 1978, Jones 1978, Mixter 1978, Robb 1981, Anderson 1982, Granberg-Michaelson 1983a, Canadian Scientific and Christian Affiliation (Guelph Chapter) 1984, Herrmann and Templeton 1985, Mooney 1985, Verhey 1985, Chamberland

1986, Erickson 1987, Kimbrell 1987, Kolkman 1987, Polman 1987a, Barnes 1989, Cook 1989, Bhagat 1990, Macer 1990, Meyer and Meyer 1991.

History of the Church's Relationship with Creation

Raven 1940, McKenzie 1952, Raven 1953, Hooykaas 1961, Sittler 1962, Ditmanson 1964, Glacken 1967, Nasr 1968, Wallace-Hadrill 1968, Glacken 1970a, b; Santmire 1970a,b; Wrightsman 1970, Macquarrie 1971, Leiss 1972, Richardson 1972, Obayashi 1973, Spring and Spring 1974, Gruner 1975, Hughes 1975, Passmore 1975, Alpers 1976, Coleman 1976, Stevenson 1976, Berkhof 1977, Worster 1977, Hargrove 1979, Squadrito 1979, Vickery 1979, Santmire 1980, Wilkinson 1980a, Yamauchi 1980, Stratman 1982, Attfield 1983a, Bamford 1983, Rigdon 1983, Thomas 1983, Rounder 1984, Thomas 1984, Abraham 1985, Berry 1985, Clarke 1985, Cohen 1985, Miles 1985, Spradley 1985, Birtel 1987, Peacocke 1987, Toulmin 1987, Granberg-Michaelson 1988c, Granberg-Michaelson 1989b, Nash 1989, Zizioulas 1989, Zizioulas 1990, Regenstein 1991.

Just Treatment of Animals

No attempt is made here to survey the literature on the broad subject of animal rights. Westermarck 1939 provides an early historical summary of Christian doctrine relating to animals in contrast to that of several Eastern religions. Aptowitzer 1926 presents a Jewish view. Examples of works addressing the just treatment of animals as part of our stewardly responsibility include Turner 1964, Kingston 1968, Linzey 1976, Morris and Fox 1978, Carpenter 1980, Griffiths 1982, Herscovici 1985, Linzey 1987, Smith (H.) 1987, Cook 1988, Kolkman 1988, McDaniel 1988, Miller 1988, Morrison 1988, Polman 1988, Ahlers 1990, Linzey 1990, Stafford 1990, Wilkinson 1990, Regenstein 1991.

Land and Agriculture

Bailey 1915, MacKay 1950, Henrey 1954-55, Hicks
1954, Brubaker 1962, Baer 1966, Hertz 1970, Davies 1974,
Westwood 1974, Bennett 1976, Austin 1977a, Berry 1977,
Brueggemann 1977, Hessel 1977, Owens 1977, Pignone 1978,
Abrecht 1978 -- part 3, Alfaro 1978, Geiger 1978, Naidoff 1978,
Walther 1978, Zylstra 1978, Everett 1979, Nelson 1979,
Brueggemann 1980, Helgeland 1980, Hough 1980, Lutz 1980,
Wang 1980, Berry 1981, Cesaretti and Commins 1981,
Freudenberger 1981, Hough 1981, Limburg 1981, Schecter
1981, Tiemstra 1981, American Lutheran Church 1982,
Davies 1982, Freudenberger 1982, Platt 1982, Schaeffer 1982,
Spaling 1982, Zylstra 1982, Gibson 1983a, Granberg-
Michaelson 1983b, Kardong 1983, Berry 1984, Brueggemann
1984, Freudenberger 1984, Hart 1984, Jackson et al. 1984,
McDonagh 1984, Montesano 1984, Nelson 1984, Stivers 1984,
Brand 1985, Bushwick 1985, Meeks 1985, Thorkelson 1985,
Anderson 1986, Benjamin 1986, Blanchard 1986,
Brueggemann 1986, de Roo 1986, Dingman 1986,
Freudenberger 1986, Heideman 1986, Herron 1986, Hultgren
1986a,b; Kanten 1986, Kraps 1986, Land Stewardship Council
of North Carolina 1986, McDonagh (S.) 1986, Mundahl 1986,
Olson 1986, Ostendorf 1986, Paddock, Paddock, and Bly
1986; Sorenson 1986, Tinker 1986, van Donkersgoed 1986,
Zinkand 1986, Aay 1987, Austin 1987c, Baltaz 1987, Brand
1987, Brubaker 1987, Brueggemann 1987, Clapp 1987, Conroy
1987b, DeWitt 1987a, Dorr 1987, Dusilek 1987, Evans 1987,
Evans and Cusack 1987, Freudenberger 1987a,b,c; Garriott
1987, Goossen 1987, Griffioen-Drenth 1987, Hart 1987,
Hawksley 1987, Helmuth 1987, Hutten 1987, Isenhart 1987,
Jackson 1987a,b; King 1987, Kline 1987, MacGillis 1987a,
McQuail 1987, Oegema 1987, Park 1987, Polman 1987b,c;
Reynolds 1987, Rybeck and Pasquariello 1987, Sikkema
1987, Stamm 1987, Tokar 1987, Vander Schaaf 1987, Weber
1987, Austin 1988b, Francis 1988, Freudenberger 1988,
Gibson 1988/89, King and Piltz 1988, McDaniel 1988,
McFarland 1988, Pfrimmer 1988, Wylie 1988, Aay 1989,
Anderson 1989a,b,c; Baltaz 1989, Birkenfeld 1989,
Brueggemann 1989, Clarke 1989, Cormack 1989, DeWitt

1989d, Freudenberger 1989a, Kolk 1989, Le-Chau 1989, Lilburne 1989, Muchina 1989, Naveh 1989, Rzadki and Hilts 1989, Schlegel 1989, Sikkema 1989, Steward 1989, Stewart-Kroeker and Kroeker 1989, Swincer 1989, van Donkersgoed 1989, Visser 1989, World Council of Churches 1989, Bhagat 1990, Brandenbarg 1990, Brubaker and Sider 1990, DeWitt 1990a,d; Freudenberger 1990, Glover 1990, Gurnett 1990, Hulteen and Jaudon 1990, Sheldon 1990, Vander Zee and Vos 1990, Van Donkersgoed 1990, Wright 1990, Adema 1991, Feddema 1991, Markus et al. 1991, Meyer and Meyer 1991, Moerman 1991, Slattery 1991, Yutzy 1991.

Lifestyle -- *see* Technology and Lifestyles

Nuclear Energy

Francis and Albrecht 1976, Hoekema 1979, American Scientific Affiliation 1980 (symposium on nuclear energy), Bube 1980, Case 1980, Cohen 1980, Ehlers 1980, Irish 1980a,b, Martin 1980, Maxey 1980, Pollard 1980a,b, Willis 1980, Aukerman 1981, Sider and Taylor 1982, Bube 1983c, Denisenko 1985, *American Baptist* 1986b,c; French 1986, Davis 1987, Hodgson 1987, McFague 1987, Meyer 1987, Sytsma 1991.

Pollution, Toxic Chemicals, Solid and Hazardous Waste

DeWolf 1970, Faramelli 1970, Hefley 1970, Widman 1970, Birch 1971, Klotz 1971b, Faramelli 1972, Guttmacher 1972, Fandrich 1973, Gillette 1973, LaBar 1974, Baer 1975, Byron 1975, Dumas 1975, Moss 1975, Freudenberger 1981, Terman (C. R.) 1982, Baer 1985, Denisenko 1985, Russell 1985, Kolkman 1986, Watson 1986, Boegman and Els 1987, Bonner 1987, Eternity 1987, Froelich 1987b, Kirkland 1987, Massey 1987, Peters (T.) 1987, Toerien 1987, Brinkman 1988, King and Piltz 1988, Ogle 1988, Bente 1989, Brubaker and Sider 1989, DeWitt 1989d, Gelderloos 1989, Gibson 1989a,b; Morud 1989e, Pater 1989, Seldman 1989, Sheldon 1989b, Squiers 1989, Bhagat 1990, DeWitt 1990a,d; Hay 1990, Head and Guerrero 1990, Lee 1990, Sheldon 1990, Sider 1990c, Badke 1991, Meyer and Meyer 1991, Sytsma 1991.

Population Issues

Fagley 1960, Brubaker 1962, Fagley 1962, Knudsen 1962, Ogg and Butcher 1962, Smalley 1962, Weaver 1962, Weaver 1966, Cobb 1970b, Christian Medical Society 1970, DeWolf 1970, Mixter 1970, Shinn 1970b, Smith 1970, Stipe 1970, Birch 1971, Finnin and Huisingh 1972, Guttmacher 1972, Mixter 1973, United Presbyterian Church (U.S.A.) 1973, Albert 1974, Bohon 1974, Carpenter 1974, Giles 1974, Gingerich 1974, Hearn 1974, Heddendorf 1974, Lewthwaite 1974, Maatman 1974, Mixter 1974, Moberg 1974, Pattison 1974, Stipe 1974, Terman 1974, Ward 1974, Yamauchi 1974, Pobee 1977, Wang 1980, Nelson 1984, Stivers 1986, Van Eeden 1987, Reynolds 1987, DeWitt 1989c,d; Bhagat 1990, DeWitt 1990a,d; Lawton 1990, Sider 1990b, Meyer and Meyer 1991, Nisly 1991, Schutz 1991, Bratton 1992.

Racial Justice and Linkage to Environmental Problems

Gibson 1989a, Head and Guerrero 1990, Lee 1990, Somplatsky-Jarman 1990.

Resources (Natural) -- *see* Energy (Non-Nuclear) and Other Natural Resources

Stewardship

Sources dealing with "stewardship" are now frequently including reference to environmental responsibility. Examples include Conrad 1954, Kantonen 1956, Thompson 1960, Peacocke 1961, Brattgard 1963, Thompson 1965, Kuhn 1970, Hendricks 1971, Smith 1971, Byron 1975, Bube 1976, Fisher 1976, Bube 1977, King 1977, Byron 1978, Jegen and Manno 1978, Swartley 1978, Walther 1978, Ehrenfeld and Bentley 1982, Hall 1982, United Presbyterian Church (U.S.A.) 1982, Yoder 1983, McDonald 1984, Bryan and Anson 1985, Ehrenfeld and Bentley 1985

(Jewish), Gibson 1985c, Hall 1985, Marshall 1985, Murphy 1985, Winn 1985, Klay 1986, Dyrness 1987, Henry 1987b, Lofgren 1987, Engle 1988, Halver 1989, Van Hoeven 1989a, Scoville 1990.

Technology and Lifestyles

Ellul 1964, White 1964, Doan 1971, Elwood 1971, Faramelli 1971, Carothers et al. 1972, Ferkiss 1972, Shinn 1972, Brown 1973, Schumacher 1973, Cobb 1974b, Ferkiss 1974, Gilkey 1974, Baer 1975, Klewin 1975, Morikawa 1975a, Baldwin 1976, Birch (C.) 1976a, Longacre 1976, Patey 1976, Bockmuhl 1977, Bye 1977, Derr 1977, Finnerty 1977, Gibson 1977b, Schumacher 1977, Sider 1977, Taylor 1977, Abrecht 1978 -- part 4, Foster 1978, Gibson 1978, Beach 1979, Kim 1979, Schumacher 1979, Schwarz 1979a, Barbour 1980, International Consultation on Simple Lifestyle 1980, Longacre 1980, Osei-Mensah 1980, Sider 1980; Sheaffer and Brand 1980, Freudenberger 1981, Foster 1981, Gibson 1981, Hessel and Wilson 1981, Sine 1981, Kroll 1982, Sider 1982, Catherwood 1983, Marshall 1983, Treece 1983, Borgmann 1984, Dorr 1984, George 1984, Baer 1985, Ballard 1985, Ramon and Bube 1985, Mooney 1985, Dammers 1986, Devadas 1986, French 1986, Gittens 1986, Monsma 1986, Shi 1986, Applegarth 1987, Freeman 1987, Froelich 1987b, Gonzalas 1987, Hogan 1987, Kolkman 1987, McDaniel (J.) 1987, Polman 1987a, Thomas 1987, Birch (C.) 1988, Aay 1989, Freudenberger 1989a, Kantzer 1989, Kroeker 1989, Maynard-Reid 1989, Morud 1989c, Neff 1989, Shuff 1989, Van Leeuwen 1989, Sider 1990b, Badke 1991. Klink 1974, rather than expressing concern for the abuses of our high technology society (which he acknowledges), sees technology as a means of reaching a "fuller and deeper relationship" in the I-Thou relationship with nature. His thinking was deeply influenced by Martin Buber. A similar position is held by Marshall 1984.

Water

Batchelor 1986, Houston 1986, Riverson 1986, Steward 1986, Froelich 1987b, Toerien 1987, DeWitt 1989d.

Wilderness

Williams 1962, Dannen 1972, Bugbee 1974, Vickery 1979, Douglas 1984, Austin 1987a, Krueger 1987b, Riggs 1987, Wharton 1987, Bratton 1988.

ENVIRONMENTAL CONCERNS WITHIN JUDAISM AND ISLAM

Other faiths associated with the Judeo-Christian tradition have also responded to the growing level of environmental concern. Although this paper deals primarily with the Christian literature, other examples will be cited to provide an understanding of the scope and nature of their concern. Included here is the Jewish tradition -- Lamm 1964-65, Hoenig 1968-69, Freudenstein 1970, Gordis 1970, Pelcovitz 1970, Vorspan 1970, Gordis 1971, Helfand 1971, Lamm 1971, Konvitz 1972, Lamm 1972, Karff 1974, Jobling 1977, Ehrenfeld and Bentley 1982, Schaffer 1982, Weiss 1983, Schwartz 1984, Clifford 1985, Ehrenfeld and Bentley 1985, Genesis Rabbah 1985, Gordis 1985, Hargrove 1986, Cohen 1990, Learner 1990, Waskow 1990, Regenstein 1991; Islam -- Zaidi 1981, Hargrove 1986, Tang 1990, Timm 1990, Regenstein 1991.

ENVIRONMENTAL CONCERNS WITHIN BUDDHISM

Davies 1987, Htun 1987, Kabilsingh 1987, Nash 1987, Dalai Lama 1987, Graef 1990, Tang 1990, Regenstein 1991.

OTHER RELIGIONS

Sebahire 1990, Tang 1990.

CHRISTIAN ORGANIZATIONS WITH A FOCUS ON CREATION

Several Christian (or interfaith) organizations have been formed to provide information on environmental matters and assist individuals who wish personally to become involved in redeeming Creation. Booklets, films and other information are available through some. For further information write to:

American Scientific Affiliation, 762 Arlington Avenue, Berkeley, CA 94707. Publishes *The ASA Newsletter.* Bimonthly. Subscription: membership. Also publishes *Perspectives on Science and Christian Faith.* Quarterly. Subscription: Membership.

Au Sable Institute of Environmental Studies, 7526 Sunset Trail N.E., Mancelona, Michigan 49659. Publishes *Au Sable Notes,* an occasional publication. Subscription: donation.

Christian Ecology Group, Secretary, Mrs. Judith Pritchard, 58 Quest Hills Road, Malvern, Worcs. WR14 1RW United Kingdom

Christian Farmers Federation of Alberta. 10766 - 97 Street, Edmonton, Alberta T5H 2M1.

Christian Farmers Federation of Ontario. 115 Woolwich Street, Guelph, Ontario N1H 3V1. Publishes, with the Christian Farmers Federation of Alberta, *Earthkeeping: A Quarterly on Faith and Agriculture.* Subscription: $15 per year.

Christian Nature Federation. CNF, 2381 Daphne Place,

Fullerton, CA 92633. Organization is described by its president as "a combination of the Audubon Society, National Wildlife Federation, and Sierra Club -- but with a Christian world view."

The Creator's Mandate, 150-207 Castlegreen Dr., Thunder Bay, Ont. P7A 7L8, Canada; (807) 767-9237. This is a bi-monthly newsletter begun in January 1990 for evangelicals focusing on issues of the environment.

Eco-justice Working Group, NCC and CRESP. Anabel Taylor Hall, Cornell University, Ithaca, NY 14853. Publishes *The Egg: A Journal of Eco-Justice.* Subscription: $7.50 per year.

Educational Concerns for Hunger Organization, Inc. (ECHO), RR2, Box 852, North Fort Meyers, FL 33903. Focus of their work is missionary agriculture.

Eleventh Commandment Fellowship. 1555 Rose Ave., Santa Rosa, CA 95407. Based on the 11th commandment: "The Earth is the Lord's and the fullness thereof; Though shalt not despoil the Earth, nor destroy the life thereon." Free brochure. Publishes *The Eleventh Commandment Newsletter.* Occasional publication. Subscription: donation.

Floresta USA, 9230 Trade Place, Suite 100, San Diego, Calif. 92126; (619) 566-6068. A Christian ministry of reforestation and responsible agriculture in the Dominican Republic with intent to expand its work to other developing countries.

Friends Committee on Unity with Nature, 7899 St. Helena Road, Santa Rosa, CA 95404. Publishes *BeFriending Creation.* Quarterly. Subscription: $15 per year.

Institute for Ecosophical Studies, c/o Moravian College, Bethlehem, PA 18018. Publishes *Ecospirit.* Subscription: donation.

Interfaith Coalition on Energy. P.O. Box 26577, Philadelphia, PA 19141. Publishes ICE Melter Newsletter.

Land Stewardship Project. 512 West Elm, Stillwater, MN 55082. Publishes *The Land Stewardship Letter.* Subscription: $15 per year.

National Catholic Rural Life Conference, 4625 Beaver
Drive, Des Moines, IA 50310. Publishes *Catholic
Rural Life.* Bi-monthly. Subscription: $20 per year.

National Impact, 100 Maryland Avenue, N.E., Washington,
D.C. 20002.

Network for Environmental and Economic Responsibility
for the UCC. 1443 Edgecliff Lane, Pasadena, CA
91107. Publishes *NEER Quarterly Newsletter.*
Subscription: contribution.

New Creation Institute. 518 S. Ave. W., Missoula, MT
59801. Stated goal is "to convert the church by its own
gospel for saving God's creation and building human wholeness."

North American Conference on Christianity and Ecology,
P.O. Box 14305, San Francisco, CA 94114. Publishes
Firmament: The Quarterly of Christian Ecology.
Subscription: $12 per year.

North American Coalition on Religion and Ecology, 5
Thomas Circle, NW, Washington, DC 20005.
Telephone 202/462-2591. Membership is $25.00 per
year. You will receive 4 issues of *Eco-letter* and two
special reports on events and resources available
from NACRE.

North Carolina Land Stewardship Council, Route 4, Box 426,
Pittsboro, NC 27312. Publishes *Cry North Carolina.*
Subscription: donation.

Wilderness Manna. P.O. Box 22233, Juneau, Alaska 99802-
2233. Publication: *Manna from the Wilderness.*
Irregular publication. Subscription: $5 per year.

CURRICULUM MATERIALS ON CREATION CARE

Christian Impact Publications. Video and filmstrip
material available on environmental stewardship
issues as well as booklets designed for Bible study
and Christian education. Inquiries should be
directed to: Christian Impact, St. Peter's Church,
Vere Street, London W1M 9HP. Telephone 071 629
3615

Eco-Justice Task Force, Committee on Social Witness
Policy, Presbyterian Church (U.S.A.). 1989.
Keeping and healing the Creation. Available from
D.M.S. Attn: Cashier, 100 Witherspoon Street,
Louisville, KY 40202-1396. 108 pages. High school
to adult.

Living the Word Series. *To Love the Earth.* St. Louis,
Missouri 63166: Christian Board of Publication, 1316
Convention Plaza, Box 179.

Mennonite Central Committee of Ontario. Will provide an
"Earth Stewardship Packet" or information on
Sunday School material. Contact: Mennonite
Central Committee Ontario, 50 Kent Ave., Kitchener,
Ontario N2G 3R1.

Outdoor Ministries Curriculum. *Creation Called to
Freedom.* The curriculum focuses on stewardship of
God's creation. Divided into five daily themes.
Available from Augsburg Fortress Publishers at 1-
800/328-4648. Elementary level and up.

Stewart, R. G. 1990. *Environmental Stewardship.* Part of the
Global Issues Bible Studies, S. Hayner and

G. Aeschliman, editors. Downers Grove, IL: Inter-Varsity Press. Six studies for individuals or groups. 48 pages.

"The Earth Is The Lord's." Indianapolis, Indiana 46206: Task Force on Christian Life Style and Ecology, Christian Church, P. O. Box 1986.

The Pelican Project: Stewardship Education for Children. Developed to teach elementary school children stewardship of creation. Includes program planning guide, video, puppet, and Erickson and Erickson 1991: *Seven Days to Care for God's World*. A small book cited in bibliography suitable to use at early grade level to teach the concept of stewardship of creation. Available from Augsburg Fortress Publishers at 1-800/328-4648.

Vacation Bible School Curriculum from Augsburg. The 1993 theme will be "Living in God's Creation." Contact Augsburg at 1-800/328-4648 for information.

THE BIBLIOGRAPHY

Aay, H. (1972, November). Confronting the ecological crisis: the kingdom of God in geographical perspective. *Vanguard*, pp. 7-27.
 (Topics):popular.

Aay, H. (1982). Images of the natural: nature as referent system in print advertising. In E. R. Squiers (Ed.), *The environmental crisis: the ethical dilemma* (pp. 187-200). Mancelona, MI: Au Sable Institute of Environmental Studies.
 (Topics):ethics.

Aay, H. (1987). Toward a just and sustainable food system. *Earthkeeping: A Quarterly on Faith and Agriculture*, 3(1), 9-10.
 (Topics):land.

Aay, H. (1989). African realities help disclose vision of development. *Earthkeeping: A Quarterly on Faith and Agriculture*, 5(2), 4-9.
 The developmental assistance given to African nations should lead to the healing of damaged and broken ecosystems.
 (Topics):popular, land, agriculture, development, technology.

Abraham, E. (1985). *Subtheme "in Christ -- hope for creation": presentation on caring for creation* (No. 19-20). Lutheran World Federation.
Focuses on the work of the Lutheran World Fellowship. Suggests that the lack of economic freedom and some of the environmental damage of the third world is due to overconsumption of resources by the industrial nations.
(Topics):popular, history, ecojustice.

Abrecht, P. (1978). *Faith, science and the future.* Geneva: Church and Society, World Council of Churches.
The book is divided into five parts: theological and ethical issues; energy; food; science and technology; and economic issues for a just, participatory, and sustainable society.
(Topics):land, book, resources, technology.

Abrecht, P. (ed.) (1980). *Reports and recommendations. In the series "Faith and science in an unjust world: report of the World Council of Churches' conference on faith, science and the future."* Philadelphia: Fortress Press.
(Topics):popular book.

Adams, J. (1982). *The solar church.* New York: The Pilgrim Press.
(Topics):solar energy.

Addinall, P. (1980-81). Walther Eichrodt and the Old Testament view of nature. *The Expository Times, 91,* 174-178.
(Topics):theology.

Adema, B. (1991). We are but sojourners: a look at the concept of land trusts. *Earthkeeping: a quarterly on faith and agriculture -- Ontario, 1*(2), 16-18.
(Topics):land.

Adkinson, L. (1982). Environmental systems modeling: toward ethical alternatives. In E. R. Squiers (Ed.), *The environmental crisis: the ethical dilemma* (pp. 201-208).

Mancelona, MI: Au Sable Institute of Environmental Studies.
(Topics):ethics.

Aeschliman, G. (1990). Earth Day: celebrate a Christian holy day. *World Christian*, 9(4), 9.
Editorial.
(Topics):popular.

Agricultural Concerns Committee (Mennonite Central Committee of Ontario), George Fast, Editor (1989). *Fragile: handle with care: an earth keeping handbook.* 36 pages.
This is a "how to" booklet on caring for Creation. Available from the Mennonite Central Committee Ontario office at 50 Kent Ave., Kitchener, ON, N2G 3R1, Canada.
(Topics):popular.

Ahlers, J. (1990). Thinking like a mountain: toward a sensible land ethic. *Christian Century*, 107(14), 433-434.
(Topics):animal rights.

Aidan, B. (1989). "Where the river flows": ecology and the orthodox liturgy. *Epiphany*, 9, 31-36.
Previous work is published under the name Arhanasius Hart. Brother Aidan is a monk at St. Elias Monastery in Wales.
(Topics):popular.

Albert, J. D. (1974). A Highly Commendable Program. *Journal of the American Scientific Affiliation*, 26(1), 15.
(Topics):population.

Albert, J. D. (1978). A journal symposium -- the recombinant DNA controversy: worthy goals and Genesis mandate outweigh dangers. *Journal of the American Scientific Affiliation*, 30(2), 80-81.
(Topics):genetic engineering.

Alfaro, J. I. (1978). The land -- stewardship. *Biblical Theology Bulletin*, 8, 51-61.

(Topics):land, theology.

Allchin, A. M. (1974). *Wholeness and transfiguration illustrated in the lives of St. Francis of Assisi and St. Seraphim of Sarov.* London: Oxford.
(Topics):popular, St. Francis of Assisi.

Allen, R. F., Ulrich, D. M., & Foti, C. P. (1973). *The works of thy hands: scripture for reflection in an age of environmental crisis.* Winona, Minn.: St. Mary's College Press.
(Topics):popular book.

Allerton, J. (1985). About a theology of conservation. *Faith and Freedom: A Journal of Progressive Religion,* 38(3), 114-126.
(Topics):theology, resources.

Allison, L. M. (1960). *Grace and nature in the theology of John Calvin.* Ph.D. diss., Union Theological Seminary in Virginia.
(Topics):theology.

Alpers, K. P. (1971). Starting points for an ecological theology: a bibliographic survey. In M. E. Marty & D. G. Peerman (Eds.), *New Theology No. 8* (pp. 292-312). New York: Macmillan.
(Topics):theology.

Alpers, K. P. (1976). Toward an environmental ethic. *Dialog,* 15(Winter), 49-55.
(Topics):history, popular, ethics.

Alvarado, R. C. (1986/87). Environmentalism and Christianity's ethic of dominion. *Journal of Christian Reconstruction,* 11(2), 201-215.
(Topics):ethics, popular.

Alves Da Silva, D. (1989, May-June). Brazil: Chico Mendez, ecologist, assassinated; interview. *Latin American Documentation,* pp. 9-17.

(Topics):popular, Chico Mendez.

American Baptist Churches in the U.S.A. Board of National Ministries (1974). Ecological wholeness and justice: the imperative of God. *Foundations*, 17(2), 133-157.
(Topics):popular, ecojustice.

American Baptist Churches in the U.S.A. General Board Reference #7004:6/77 (1986a). Energy. *American Baptist Quarterly*, 5(2,3), 108-112.
(Topics):energy.

American Baptist Churches in the U.S.A. General Board Reference #8053:9/82 (1986b). The disposal of hazardous and radioactive wastes. *American Baptist Quarterly*, 5(2,3), 203-204.
(Topics):hazardous waste, nuclear.

American Baptist Churches in the U.S.A. General Board Reference #8097:12/82 (1986c). Nuclear power: seeking rational solutions. *American Baptist Quarterly*, 5(2,3), 259-263.
(Topics):energy, nuclear.

American Baptist Churches in the U.S.A. General Board Reference #8106:6/83 (1986d). In support of the Law of the Sea Treaty. *American Baptist Quarterly*, 5(2,3), 248-249.
(Topics):popular.

American Baptist Churches in the U.S.A. General Board Reference #8114:3/83 (1986e). Environmental concerns. *American Baptist Quarterly*, 5(2,3), 213-214.
(Topics):popular.

American Lutheran Church (1982). *The land: statements and actions of the American Lutheran Church (1978-1982) dealing with the land and those who tend it.* Minneapolis: Augsburg.
(Topics):land.

American Scientific Affiliation (1969). *Journal of the American Scientific Affiliation*, 21(2), 33-47.
Thematic issue dealing with environmental concerns.
(Topics):popular.

American Scientific Affiliation (1978). A journal symposium: the recombinant DNA controversy. *Journal of the American Scientific Affiliation*, 30(2), 73-81.
A group of several short papers examining the "recombinant DNA controversy."
(Topics):genetic engineering.

American Scientific Affiliation (1980). Symposium on nuclear energy. *Journal of the American Scientific Affiliation*, 32(2), 65-114.
Thematic issue focusing on nuclear energy issues.
(Topics):nuclear energy.

Ancil, R. E. (1990). Man and his environment: a creationist perspective. *Creation, Social Science and Humanities quarterly*, 12(Summer), 19-22.
(Topics):popular.

Anderson, B. W. (1955). The Earth Is the Lord's. *Interpretation*, 9(1), 3-20.
(Topics):theology.

Anderson, B. W. (1967). *Creation versus chaos*. New York: Association Press.
(Topics):theology.

Anderson, B. W. (1975). Human dominion over nature. In M. Ward (Ed.), *Biblical studies in contemporary thought* (pp. 27-45). Somerville, Mass: Greeno, Hadden & Co.
(Topics):theology.

Anderson, B. W. (1983). Creation and ecology. *American Journal of Theology and Philosophy*, 4(1), 14-30.
(Topics):theology.

Anderson, B. W. (1984). *Creation in the Old Testament.* Philadelphia: Fortress Press.
(Topics):theology.

Anderson, B. W. (1987). *Creation versus chaos.* Philadelphia: Fortress Press.
(Topics):theology.

Anderson, F. (1989a). In search of a covenant economy: the Catholic bishops of Ontario issue a call for economic justice for farmers and for urban understanding. *Earthkeeping: A Quarterly on Faith and Agriculture,* 5(4/5), 6-8.
(Topics):land, agriculture.

Anderson, F. (1989b). Keeping land in perspective. *Earthkeeping: A Quarterly on Faith and Agriculture,* 5(4/5), 3.
Editorial.
(Topics):land.

Anderson, F. (1989c). Stewardship: opportunities and challenges. *Earthkeeping: A Quarterly on Faith and Agriculture,* 5(1), 19-20.
(Topics):agriculture, land.

Anderson, J. K. (1982). *Genetic engineering.* Grand Rapids, MI: Academic Books, Zondervan.
(Topics):genetic engineering.

Anderson, T. (1986). Indigenous peoples, the land, and self-government. In B. W. Harrison, R. L. Stivers, & R. H. Stone (Eds.), *The public vocation of Christian ethics* (pp. 159-178). New York: The Pilgrim Press.
(Topics):ethics, land.

Applegarth, A. (1987). Should Christians live more simply? *Virtue,* 10(3), 62-64.
(Topics):lifestyle.

Aptowitzer, V. (1926). The rewarding and punishing of animals and inanimate objects: on the aggadic view of the world. *Hebrew Union College Annual*, 3, 117-155.
Jewish.
(Topics):animal rights, theology.

Armerding, C. E. (1973). Biblical perspectives on the ecology crisis. *Journal of the American Scientific Affiliation*, 25(1), 4-9.
(Topics):popular.

Armstrong, E. A. (1973). *Saint Francis: nature mystic*. Berkeley, Calif.: University of California Press.
(Topics):Saint Francis of Assisi.

Arnold, E. (1987). The implications of eco-justice. In F. W. Krueger (Ed.), *Christian ecology: building an environmental ethic for the twenty-first century* (p. 87). North Webster, Indiana: The North American Conference on Christianity and Ecology.
(Topics):popular, ecojustice.

Arnold, E. B. (1988/89). The limitations of enlightened self-interest. *The Egg: A Journal of Eco-Justice*, 8(4), 6-8.
(Topics):popular.

Arvill, R. (1969). *Man and environment*. London: Penguin Books.
A brief note to man's dominion over nature appears on page 287 with reference to Genesis and Hebrews.
(Topics):popular book.

Aschliman, K. A. (1987). Parenting for peace and justice and its relationship to the environement. In F. W. Krueger (Ed.), *Christian ecology: building an environmental ethic for the twenty-first century* (p. 79). North Webster, Indiana: The North American Conference on Christianity and Ecology.
(Topics):popular.

Asian Theologians (1977). Man and nature. *3cc Bul Miss R,* 1, 19-21.
Workshop Statement by Asian Theologians in Manila. The journal abbreviation is as it appeared in *Religion Index One:* Periodicals, Vol. 13:336.
(Topics):popular.

Asselin, D. T. (1954). The notion of dominion in Genesis 1-3. *Catholic Biblical Quarterly,* 16, 277-294.
(Topics):theology.

Attfield, R. (1983a). Christian attitudes to nature. *Journal of the History of Ideas,* 44(3), 369-386.
(Topics):history.

Attfield, R. (1983b). *The Ethics of Environmental Concern.* New York: Columbia University Press.
(Topics):ethics, popular book.

Attfield, R. (1983c). Western traditions and environmental ethics. In R. Elliot & A. Gare (Eds.), *Environmental philosophy: a collection of readings* (pp. 201-230). University Park: Pennsylvania State University Press.
(Topics):ethics.

Aukerman, D. (1981). *Darkening valley: a biblical perspective on nuclear war.* New York: Seabury Press.
(Topics):nuclear.

Aukerman, D. (1983). Forfeited dominion. *Epiphany,* 3(3), 18-20.
(Topics):popular.

Austin, R. C. (1977a). Three axioms for land use. *The Christian Century,* 94(32), 910-915.
(Topics):popular, land.

Austin, R. C. (1977b). Toward environmental theology. *The Drew Gateway,* 48(2), 1-14.
(Topics):theology, popular.

Austin, R. C. (1987a). *Baptized into wilderness: a Christian perspective on John Muir.* Atlanta, GA: John Knox Press.
(Topics):book, wilderness.

Austin, R. C. (1987b). Beyond stewardship to relationship: John Muir's challenge to Christianity. In F. W. Krueger (Ed.), *Christian ecology: building an environmental ethic for the twenty-first century* (p. 51). North Webster, Indiana: The North American Conference on Christianity and Ecology.
(Topics):popular.

Austin, R. C. (1987c). Rights for life: rebuilding human relationships with land. In B. F. Evans & G. D. Cusack (Eds.), *Theology of the land* (pp. 103-126). Collegeville, MN: The Liturgical Press.
(Topics):land, theology.

Austin, R. C. (1988a). *Beauty of the Lord.* Atlanta, GA: John Knox Press.
(Topics):book.

Austin, R. C. (1988b). *Hope for the land: nature in the Bible.* Atlanta, GA: John Knox Press.
(Topics):land, book.

Axel, L. E., Peden, W. C., & eds. (1983). Technology, culture, and environment. *American Journal of Theology and Philosophy,* 4(1), 3-48.
Thematic issue.
(Topics):theology.

Badke, W. B. (1991). *Project Earth: preserving the world God created.* Portland, OR: Multnomah Press.
(Topics):pollution, lifestyle, popular book.

Baer, R. A., Jr. (1966, Oct. 12). Land misuse: a theological concern. *Christian Century,* pp. 1239-1241.
(Topics):land.

Baer, R. A., Jr. (1969). Conservation: an arena for the church's concern. *Christian Century*, 86(2), 40-43.
(Topics):popular, resources.

Baer, R. A., Jr. (1970). Christian responsibility in man's relation to nature. *The Lutheran Forum*, 4, 6-7.
(Topics):popular.

Baer, R. A., Jr. (1971). Ecology, religion, and the American dream. *American Ecclesiastical Review*, 165, 43-59.
(Topics):ethics, popular.

Baer, R. A., Jr. (1975). Poverty, pollution, and the power of the gospel. *Engage / Social Action*, 3, 49-57.
(Topics):lifestyle, pollution.

Baer, R. A., Jr. (1977). Higher education, the church and environmental values. *Natural Resources Journal*, 17, 477-491.
(Topics):popular, education.

Baer, R. A., Jr. (1985). Poverty, pollution, and the power of the gospel. In N. C. Murphy (Ed.), *Teaching and preaching stewardship: an anthology* (pp. 223-231). New York: Commission on Stewardship, National Council of the Churches of Christ in the U.S.A.
(Topics):pollution, lifestyle.

Bagby, R. (1987). Eco-feminism and the church: part three: caring for the earth. In F. W. Krueger (Ed.), *Christian ecology: building an environmental ethic for the twenty-first century* (p. 100). North Webster, Indiana: The North American Conference on Christianity and Ecology.
(Topics):eco-feminism, popular.

Bailey, L. H. (1915). *The holy earth*. New York: Reprinted in cooperation with the author by The Christian Rural Fellowship in 1946.
(Topics):book, agriculture, land.

Baker, J. M., and Bradbury, R., (eds.) (1990). The state of the ark. *The Modern Churchman*, 32(2), 1-55.
Christianity and the environment conference organized by Modern Churchpeople's Union on July 18 - 21, 1989.
(Topics):popular.

Balasuriya, T. (1984). *Planetary theology*. London: SCM Press.
(Topics):theology.

Baldwin, S. (1976). A case against waste and other excesses. *Christianity Today*, 20(21), 1066-1070.
(Topics):popular, lifestyle.

Ballard, P. H. (1985). Conservation in a Christian context. *The Baptist Quarterly*, 31, 23-38.
(Topics):popular.

Balmer, R. (1989). California dreamin'. *The Reformed Journal*, 39(10), 7-8.
Comments on the deteriorating condition of the California environment.
(Topics):popular.

Baltaz, D. (1987). Jesuit farm project: searching for spirituality and justice. *Earthkeeping: A Quarterly on Faith and Agriculture*, 3(4), 14-16.
(Topics):agriculture, land.

Baltaz, D. (1989). Farming as a spiritual pursuit: the Jesuit farm project examines the non-economic side of farming. *Earthkeeping: A Quarterly on Faith and Agriculture*, 5(4/5), 4-6.
(Topics):agriculture, farming.

Bamford, C. (1983). Ecology and holiness: the heritage of Celtic Christianity. *Epiphany*, 3(3), 66-78.
(Topics):history.

Bamford, C. (1984). Daily bread. *Parabola: The Magazine of Myth and Tradition*, 9(4), 56-63.
(Topics):theology.

Barbieri, C. (1987). Spiritualities in environmental education: part two: good planets are hard to come by: educating about the eleventh commandment and the emergence of ecological intrinsic values. In F. W. Krueger (Ed.), *Christian ecology: building an environmental ethic for the twenty-first century* (pp. 81-82). North Webster, Indiana: The North American Conference on Christianity and Ecology.
(Topics):popular, education.

Barbour, I. (1980). *Technology, environment, and human values*. New York: Praeger Publishers.
(Topics):lifestyle, technology, book.

Barbour, I. G. (1966). *Issues in science and religion*. Englewood Cliffs, N J: Prentice-Hall, Inc.
See chapter 13, "God and Nature."
(Topics):theology.

Barbour, I. G. (Ed.) (1968). *Science and religion: new prespectives on the dialogue.* New York: Harper & Row, Inc.
(Topics):theology.

Barbour, I. G. (1970). An ecological ethic. *Christian Century*, 87, 1180-1184.
(Topics):popular, ethics.

Barbour, I. G. (Ed.) (1972). *Earth might be fair.* Englewood Cliffs, NJ: Prentice-Hall, Inc.
(Topics):book, theology, ethics.

Barbour, I. G. (Ed.) (1973). *Western man and environmental ethics.* Reading, MA.: Addison-Wesley.
(Topics):ethics.

Barbour, I. G. (1979). Justice, participation and sustainability at MIT. *The Ecumenical Review*, 31, 380-387.

(Topics):popular, ethics.

Barnes, P. (1989). Biotechnology: separation from creation? *Earthkeeping: A Quarterly on Faith and Agriculture*, 5(3), 17-18.
(Topics):genetic engineering.

Barnett, T. L., and Flora, S. R. (1982a). *Christian outdoor education: a program for Christian schools and camps.* Duluth, Minn.: Camping Guideposts.
(Topics):education.

Barnett, T. L., and Flora, S. R. (1982b). *Exploring God's web of life: a handbook for Christian outdoor education.* Duluth, Minn.: Camping Guideposts.
(Topics):education.

Barnette, H. H. (1972a). *The church and the ecological crisis.* Grand Rapids: Eerdmans Pub. Co.
An important relatively early work for the general reader.
(Topics):book, theology, ethics.

Barnette, H. H. (1972b). Toward an ecological ethic. *Review and Expositor*, 69(1), 23-35.
(Topics):ethics.

Barney, G. O. (1990) [untitled piece accompanying the paper by van Drimmelen 1990]. *One World*, no. 161(Dec.):19.
Barney was author of *The Global 2000 Report*. He is a layman in the Evangelical Lutheran Church in America and on the staff of the Institute for 21st Century Studies.
(Topics):popular.

Barney, G. O. (1984). The future of the creation: the central challenge for theologians. *Word & World*, 4(4), 422-429.
(Topics):popular.

Barr, J. (1968-69). The images of God in the book of Genesis. *Bulletin of the John Rylands University Library*, 51, 11-26.

(Topics):theology.

Barr, J. (1972). Man and nature: the ecological controversy and the Old Testament. *Bulletin of the John Rylands University Library*, 55(1), 9-32.
(Topics):theology.

Basney, L. (1973). Ecology and the scriptural concept of the master. *Christian Scholar's Review*, 3(1), 49-50.
(Topics):popular.

Batchelor, S. (1986, July-September). Windpumps: an answer for Africa. *Together: A Journal of World Vision International*, pp. 19-20.
(Topics):water, energy, wind.

Bauckham, R. (1986). First steps to a theology of nature. *The Evangelical Quarterly*, 58(3), 229-244.
(Topics):theology.

Bauman, S. (1986). Virtue: a foundation in the heart. *Epiphany*, 6(4), 26-29.
   A call for virtue within the Church and humanity that it might extend to include the fullness of creation.
(Topics):popular.

Beach, W. (1979). *The wheel and the cross: a Christian response to the technological revolution*. Atlanta: John Knox Press.
(Topics):technology, book.

Beisner, E. C. (1990). *Prospects for growth: a biblical view of population, resources, and the future*. Westchester, Ill.: Crossway Books.
   A cornucopian view of the environment using similar arguments to those of economist Julian Simon in the secular work *The Economics of Population Growth*.
(Topics):economics, cornucopian.

Benjamin, P. (1986). *The theology of the land in a book of Joshua*. Ph.D., Lutheran School of Theology at Chicago.
(Topics):land, theology.

Benjamin, W. W. (1979). A challenge to the eco-doomsters. *The Christian Century*, 96(10), 252-272.
"Garrett Hardin and the lifeboat moralists fail to see the connection between affluence in the U. S. and starvation in Third World countries."
(Topics):popular.

Bennett, J. B. (1974). Nature -- God's body: a Whiteheadian perspective. *Philosophy Today*, 18, 248-254.
(Topics):theology.

Bennett, J. B. (1976, Summer). A context for the land ethic. *Philosophy Today*, pp. 124-133.
Discusses Whitehead's process theology as it relates to the land ethic.
(Topics):ethics, land.

Bennett, J. B. (1977). On responding to Lynn White: ecology and Christianity. *Ohio Journal of Religious Studies*, 5(1), 71-77.
(Topics):theology.

Bente, P. F. (1989). Becoming a responsible entity in God's creation. *Lutheran Theological Seminary Bulletin*, 69(3), 49-56.
(Topics):resources, pollution.

Berkhof, H. (1977). God in nature and history. In C. T. McIntire (Ed.), *God, history, and historians: modern Christian views of history*. New York: Oxford University Press.
A Faith and Order paper of the World Council of Churches.
(Topics):theology, history.

Berry, J. F. (1987). The implications of creation spirituality: part three: Earth technologies. In F. W. Krueger (Ed.), *Christian ecology: building an environmental ethic for the twenty-first century* (p. 86). North Webster, Indiana: The North American Conference on Christianity and Ecology.
(Topics):creation spirituality, popular.

Berry, R. J. (1972). *Ecology and Ethics*. Downers Grove, IL: Inter-Varsity Press.
(Topics):ethics.

Berry, T. (1980). Management: the managerial ethos and the future of planet Earth. *Teilhard Studies*, 3, 1-14.
(Topics):popular.

Berry, T. (1985). Wonderworld as wasteworld: the earth in deficit. *Cross Currents*, 35, 408-422.
(Topics):history, economics.

Berry, T. (1987a). The implications of creation spirituality: part one: overview. In F. W. Krueger (Ed.), *Christian ecology: building an environmental ethic for the twenty-first century* (p. 85). North Webster, Indiana: The North American Conference on Christianity and Ecology.
(Topics):creation spirituality, popular.

Berry, T. (1987b). The spirituality of the Earth. In F. W. Krueger (Ed.), *Christian ecology: building an environmental ethic for the twenty-first century* (p. 21). North Webster, Indiana: The North American Conference on Christianity and Ecology.
(Topics):popular, creation spirituality.

Berry, T. (1988a). *The dream of the Earth.* San Francisco: Sierra Club Books.
(Topics):creation spirituality, book.

Berry, T. (1988b). The Earth community: we must be clear about what happens when we destroy the living forms of this planet. *Christian Social Action*, 1, 11-13.

(Topics):biodiversity.

Berry, W. (1972). *A continuous harmony.* San Diego: Harcourt, Brace, Jovanovich Publishers.
(Topics):popular book.

Berry, W. (1977). *The unsettling of America.* San Francisco: Sierra Club Book.
(Topics):land.

Berry, W. (1981). *The gift of good land: further essays, cultural and agricultural.* San Francisco: North Point Press.
(Topics):land, popular book, agriculture.

Berry, W. (1984). Two economies. *Review and Expositor,* 81, 209-223.
(Topics):popular, land, economics.

Berry, W. (1987). God and country. In F. W. Krueger (Ed.), *Christian ecology: building an environmental ethic for the twenty-first century* (pp. 15-17). North Webster, Indiana: The North American Conference on Christianity and Ecology.
(Topics):popular.

Berry, W. (1988). God and Country. *Firmament: The Quarterly of Christian Ecology,* 1(1), 10-13.
(Topics):popular.

Bhagat, S. P. (1990). *Creation in crisis: responding to God's covenant.* Elgin, Ill.: Brethren Press.
  Foreword is by Calvin B. DeWitt of Au Sable Institute.
  (Topics):biblical perspectives, pollution, hazardous waste, resources, land, population, biotechnology, economics.

Bindewald, A. (1977). *Spiritual living in a technological environment.* M.A., Duquesne University.
(Topics):theology.

Birch, B. C. (1978). Energy ethics reaches the church's agenda. *The Christian Century*, 95(35), 1034-1038.
(Topics):energy, ethics.

Birch, B. C. (1988). Nature, humanity, and biblical theology: observations toward a relational theology of nature. In W. Granberg-Michaelson (Ed.), *Ecology and life: accepting our environmental responsibility* (pp. 143-150). Waco, TX: Word Books.
(Topics):theology.

Birch, B. C., & Rasmussen, L. L. (1978). *The Predicament of the Prosperous*. Philadelphia: Westminster Press.
(Topics):resources, popular book.

Birch, B. C., & Rasmussen, L. L. (1985). These all look to thee: a relational theology of nature. *Engage/Social Action*, 13(5), 33-39.
(Topics):popular, theology.

Birch, C. (1965). *Nature and God*. Philadelphia: Westminster Press.
(Topics):theology.

Birch, C. (1971). The global environment, responsible choice and social justice. *The Ecumenical Review*, 23(4), 438-442.
Approved by Executive Committee of the World Council of Churches.
(Topics):popular, resources, population, pollution.

Birch, C. (1975). *Confronting the future: Australia and the world: the next hundred years*. London: Penguin Books.
Excellent review of world problems written for a secular audience. Has some references to Christianity and environmental responsibility.
(Topics):popular book.

Birch, C. (1976a). Creation, technology and human survival: called to replenish the Earth. *The Ecumenical Review*, 28, 66-79.
    (Topics):popular, technology, appropriate technology.

Birch, C. (1976b). Ecological Liberation. *Engage / Social Action*, 4, 49-52.
    (Topics):popular.

Birch, C. (1978). *Faith, science and the future.* Geneva: World Council of Churches.
    (Topics):book.

Birch, C. (1979). Nature, God and humanity in ecological perspective. *Christianity and Crisis: A Christian Journal of Opinion*, 39, 259-266.
    (Topics):popular.

Birch, C. (1980). Nature, humanity and God in ecological perspective. In R. L. Shinn (Ed.), *Faith and science in an unjust world. Report of the World Council of Churches Conference on Faith, Science, and the Future, Vol. 1: Plenary Presentations.* Philadelphia: Fortress Press.
    (Topics):theology, popular.

Birch, C. (1988). The scientific-environmental crisis: where do the churches stand? *The Ecumenical Review*, 40, 185-193.
    (Topics):popular, technology.

Birch, C., & Cobb, J. B., Jr. (1981). *The liberation of life.* Cambridge: Cambridge University Press.
    (Topics):theology.

Birkenfeld, D. L. (1989). Land: a place where justice, peace and creation meet. *International Review of Mission*, 78(April), 155-161.
    (Topics):land.

Birtel, F. T. (Ed.) (1987). *Religion, science, and public policy*. New York: The Crossroad Publishing Company.
(Topics):history, book, theology.

Black, J. (1970). *The dominion of man: the search for ecological responsibility*. Edinburgh: The University Press.
(Topics):book.

Blackburn, J. (1972). *The Earth Is the Lord's?* Waco, Texas: Word Books.
(Topics):book.

Blaiklock, E. M. (1971). Plague Upon Us. *Eternity*, 22(4), 25.
(Topics):popular.

Blair, I. (1983). Energy and environment: the ecological debate. In J. R. W. Stott (Ed.), *The Year 2000*. Downers Grove, IL: Inter-Varsity Press.
(Topics):energy.

Blanchard, W. M., Jr. (1986). *Changing hermeneutical perspectives on "the land" in biblical theology*. Ph.D., The Southern Baptist Theological Seminary.
(Topics):land, theology.

Blewett, J. (1987). The implications of creation spirituality: part two: Holy Cross Center and the stations of the cosmic Earth. In F. W. Krueger (Ed.), *Christian ecology: building an environmental ethic for the twenty-first century* (pp. 85-86). North Webster, Indiana: The North American Conference on Christianity and Ecology.
(Topics):creation spirituality, popular.

Blewett, J. (1990). The greening of Catholic social thought? *Pro Mundi Vita Studies*, 13(February), 27-35.
(Topics):popular.

Blidstein, G. J. (1966). Man and nature in the sabbatical year. *Tradition*, 9(4), 48-55.
(Topics):theology.

Bliss, B. (1972). Ecology -- So What? *U.S. Catholic and Jubilee*, 37, 32-38.
(Topics):popular.

Blocher, H. (1984). *In the beginning: the opening chapters of Genesis*. Downers Grove, IL: Inter-Varsity Press.
(Topics):theology.

Bloomquist, K. L. (1989). Creation domination and the environment. *Lutheran Theological Seminary Bulletin*, 69(3), 27-31.
(Topics):theology.

Bockmuhl, K. (1975). Destroyer or provider. *Christianity Today*, 19(18), 911-912.
(Topics):popular.

Bockmuhl, K. (1977). *Conservation and lifestyle* (B. N. Kaye, Trans.). Bramcote, Notts.: Grove Books.
(Topics):lifestyle, theology.

Bodner, J. (Ed.) (1984). *Taking charge of our lives: living responsibly in the world*. San Francisco: Harper and Row.
(Topics):book.

Bodo, M. (1984). *The way of St. Francis*. Garden City, N. Y.: Doubleday.
(Topics):book, St. Francis of Assisi.

Boegman, N., & Els, C. J. (1987). Air pollution: Is it serious? In W. S. Vorster (Ed.), *Are We Killing God's Earth?* (pp. 89-99). Pretoria, South Africa: University of South Africa.
(Topics):air pollution.

Boettcher, F. N. (1978). *Judeo-Christian resources for valuing our environment*. D. Min., Lutheran School of Theology at Chicago.
(Topics):theology.

Bonhoeffer, D. (1959). *Creation and fall: a theological interpretation of Genesis 1-3.* New York: Macmillan. Translation by J.C. Fletcher of 1937 German edition. (Topics):theology.

Bohon, R. L. (1974). No Compulsory Control. *Journal of the American Scientific Affiliation,* 26(1), 15-16. (Topics):population.

Bonifazi, C. (1967). *A theology of things: a study of man in his physical environment.* Philadelphia: J. B. Lippincott Co. (Topics):theology.

Bonifazi, C. (1970). Biblical roots of an ecologic conscience. In M. Hamilton (Ed.), *This little planet* (pp. 203-233). New York: Charles Scribner's Sons. (Topics):theology.

Bonifazi, C. (1978). *The soul of the world: an account of the inwardness of things.* Lanham, MD: University Press of America. (Topics):theology.

Bonner, P. (1987). Who will clean up America's deadly leftovers? *Eternity,* 38(7/8), 10-12. (Topics):hazardous waste.

Borgmann, A. (1984). Prospects for the theology of technology. In C. Mitcham & J. Grote (Eds.), *Theology and technology.* Lanham, MD: University Press of America. (Topics):theology, technology.

Boulton, W. G. (1990, April 25). The thoroughly modern mysticism of Matthew Fox. *The Christian Century,* pp. 428-432. (Topics):creation spirituality.

Boyd, G. N. (1975). Schleiermacher: On Relating Man and Nature. *Encounter,* 36, 10-19.

(Topics):popular.

Braaten, C. E. (1972). *Christ and counter-Christ: apocalyptic themes in theology and culture.* Philadelphia: Fortress Press.
See especially Chapter 8.
(Topics):theology.

Braaten, C. E. (1974a). Caring for the future: where ethics and ecology meet. *Zygon*, 9, 311-322.
(Topics):ethics, theology.

Braaten, C. E. (1974b). *Eschatology and ethics.* Minneapolis: Augsburg Publishing House.
Especially chapter 12 -- "Caring for the Future: Where Ethics and Ecology Meet."
(Topics):ethics, theology.

Brand, P. (1985). A handful of mud. *Christianity Today*, 29(7), 25-31.
(Topics):popular, land.

Brand, P. W. (1987). "A handful of mud": a personal history of my love for the soil. In W. Granberg-Michaelson (Ed.), *Tending the garden: essays on the gospel and the Earth* (pp. 136-150). Grand Rapids: Eerdmans Publishing Company.
(Topics):popular, land.

Brandenbarg, G. (1990). The Church and family farming in West Africa: a woman's perspective. *Earthkeeping: a quarterly on faith and agriculture*, 6(2):10-11.
(Topics):land, agriculture.

Brattgard, H. (1963). *God's stewards: a theological study of the principles and practices of stewardship* (G. J. Lund, Trans.). Minneapolis: Augsburg Publishing House.
(Topics):stewardship, theology.

Bratton, S. P. (1983). The eco-theology of James Watt. *Environmental Ethics*, 5(3), 225-236.
(Topics):ethics.

Bratton, S. P. (1984). Christian ecotheology and the Old Testament. *Environmental Ethics*, 6(3), 195-209.
(Topics):ethics, theology.

Bratton, S. P. (1986). Guest editorial: manager reflects on new environmental ethics program at University of Georgia. *Restoration & Management Notes*, 4(1), 3-4.
(Topics):ethics.

Bratton, S. P. (1988). The original desert solitaire: early Christian monasticism and wilderness. *Environmental Ethics*, 10(1), 31-53.
(Topics):ethics, wilderness.

Bratton, S. P. (1990). Teaching environmental ethics from a theological perspective. *Religious Education*, 85(1), 25-33.
(Topics):ethics.

Bratton, S. P. (1992). *Six billion and more: human population regulation and Christian ethics.* Louisville, KY: John Knox/Westminister Press.
    A major work on the ethics of population growth from a Christian perspective. Scheduled to be published in the spring of 1992. Foreword by Calvin B. DeWitt of Au Sable Instititute.
(Topics):ethics, population.

Bridger, F. (1990). Ecology and eschatology: a neglected dimension. *Tyndale Bulletin*, 41(November), 290-301.
(Topics):theology.

Bring, R. (1964). The gospel of the new creation. *Dialog*, 3, 274-282.
(Topics):theology.

Brinkman, M. (1988). The Christian faith as environmental pollution? *Exchange*, 17, 36-47.
(Topics):popular, pollution.

Brisbin, I. L. (1979). The principles of ecology as a frame of reference for ethical challenges: towards the development of an ecological theology. *Georgia Journal of Science*, 37, 21-34.
(Topics):theology, ethics.

Brockway, A. R. (1973). Toward a theology of the natural world. *Engage/Social Action*, 1, 21-30.
(Topics):theology.

Brockway, A. R. (1976). A theology of the natural world. *Engage/Social Action*, 4, 35-42.
(Topics):theology.

Brow, R. (1989). The taming of a New Age prophet. *Christianity Today*, 33(9), 28-30.
(Topics):Matthew Fox, Creation Spirituality.

Brown, J. P. (1971). Million-year plan. *Religious Education*, 66(1), 49-55.
(Topics):popular.

Brown, R. M. (1973). *Frontiers for the church today*. New York: Oxford University Press.
(Topics):technology.

Brown, S. (1987). Ecological programs for the local church. In F. W. Krueger (Ed.), *Christian ecology: building an environmental ethic for the twenty-first century* (pp. 97-98). North Webster, Indiana: The North American Conference on Christianity and Ecology.
(Topics):popular.

Brubaker, D., & Sider, R. J. (1989, July/August). The right to breath clean air. *ESA Advocate*, pp. 1-3,16.
(Topics):air pollution.

Brubaker, D., & Sider, R. J. (1990). The farm bill: not just for farmers. *ESA Advocate*, 12(5):1-4.
(Topics):agriculture.

Brubaker, K. K. (1962). The balance of food and population. *Journal of the American Scientific Affiliation*, 14(1), 2-7.
(Topics):population, agriculture.

Brubaker, K. K. (1987). Sustainable agriculture and the church. In F. W. Krueger (Ed.), *Christian ecology: building an environmental ethic for the twenty-first century* (p. 48). North Webster, Indiana: The North American Conference on Christianity and Ecology.
(Topics):land, agriculture.

Bruce, F. F. (1983). The Bible and the environment. In M. Inch & R. Youngblood (Eds.), *The living and active word of God: studies in honor of Samuel J. Schultz*. Winona Lake, IN: Eisenbrauns.
(Topics):theology.

Brueggemann, W. (1969, Sept. 10). King in the kingdom of things. *The Christian Century*, pp. 1165-1166.
A short article examining the ecological/theological meaning of Genesis 1:28.
(Topics):popular.

Brueggemann, W. (1970). Kings should know better. *Eternity*, 21(5), 14ff.
(Topics):popular.

Brueggemann, W. (1977). *The land: place as gift, promise, and challenge in biblical faith: overtures to biblical theology*. Philadelphia: Fortress Press.
(Topics):land, theology.

Brueggemann, W. (1980). On land losing and land receiving. *Dialog*, 19, 166-173.
(Topics):land, theology.

Brueggemann, W. (1982). *Genesis: A Bible commentary for teaching and preaching.* Atlanta: John Knox Press.
(Topics):theology.

Brueggemann, W. (1983). *Interpretation: Genesis.* Atlanta: John Knox Press.
(Topics):theology.

Brueggemann, W. (1984). Theses on land in the Bible. In Erets (Ed.), *The church and Appalachian land issues.* Amesville, Ohio: The Coalition for Appalachian Ministry.
(Topics):land, theology.

Brueggemann, W. (1986). "The Earth is the Lord's": a theology of Earth and land. *Sojourners,* 15(9), 28-32.
(Topics):land, theology.

Brueggemann, W. (1987). Land: fertility and justice. In B. F. Evans & G. D. Cusack (Eds.), *Theology of the land* (pp. 41-68). Collegeville, MN: The Liturgical Press.
(Topics):land, theology.

Brueggemann, W. (1989). The land and our urban appetites. *Perspectives: A Journal of Reformed Thought,* 4(2), 9-13.
Paper presented at Madison, Wis., conference sponsored by NACCE and World Alliance of Reformed Churches.
(Topics):land.

Brunner, E. (1952). *The Christian doctrine of creation and redemption.* London: Lutterworth Press.
(Topics):theology.

Bryan, R., & Anson, W. (1985, June). The stewardship of earth. *Quaker Life,* pp. 12-17.
(Topics):popular, stewardship.

Bryant, A. (1987). Christian ecology in Europe: part one: the British perspective. In F. W. Krueger (Ed.), *Christian ecol-*

ogy: building an environmental ethic for the twenty-first century (p. 67). North Webster, Indiana: The North American Conference on Christianity and Ecology.
(Topics):popular.

Bryce-Smith, D. (1977). Ecology, theology, and humanism. *Zygon*, 12, 212-231.
(Topics):popular.

Bube, R. H. (1974, July-August). Tomorrow's energy sources: a summary for laymen. *The Reformed Journal*, pp. 21-25.
(Topics):energy.

Bube, R. H. (1976). A new consciousness: energy and Christian stewardship. *Journal of the American Scientific Affiliation, Supplement,* 1, 8.
(Topics):energy, stewardship.

Bube, R. H. (1977). A Christian affirmation on the stewardship of natural resources. *Journal of the American Scientific Affiliation,* 29(3), 97-98.
(Topics):resources, stewardship.

Bube, R. H. (1978a, June 14). Get energy off the back burner. *Eternity*, pp. 14ff.
(Topics):energy.

Bube, R. H. (1978b). A journal symposium -- the recombinant DNA controversy: avoid simplistic thinking. *Journal of the American Scientific Affiliation,* 30(2), 78-79.
(Topics):genetic engineering.

Bube, R. H. (1980). How simple life would be if only things weren't so complicated! *Journal of the American Scientific Affiliation,* 32(2), 65-69.
(Topics):nuclear.

Bube, R. H. (1983a). Energy and the environment (A) is energy a Christian issue? *Journal of the American Scientific Affiliation*, 35(1), 33-36.
(Topics):energy.

Bube, R. H. (1983b). Energy and the environment (B) barriers to responsibility. *Journal of the American Scientific Affiliation*, 35(2), 92-100.
(Topics):energy.

Bube, R. H. (1983c). Energy and the environment (C) Christian concerns on nuclear energy and nuclear warfare. *Journal of the American Scientific Affiliation*, 35(3), 168-174.
(Topics):nuclear.

Bugbee, H. G. (1974). Wilderness in America. *Journal of the American Academy of Religion*, 42, 614-620.
(Topics):popular, wilderness.

Buitendag, J. (1986). *Creation and ecology: a systematic inquiry into the theological understanding of reality.* D.D. University of Pretoria (South Africa)
Afrikaans text.
(Topics):theology.

Bullock, W. L. (1969). The coming catastrophes: causes and remedies. *Journal of the American Scientific Affiliation*, 21(3), 84-87.
(Topics):popular.

Bullock, W. L. (1971). Ecology and apocalypse. *Christianity Today*, 15(15), 700-704.
Review of four books: *Pollution and Death of Man: the Christian View of Ecology* by Francis Shaeffer; *Brother Earth* by H. Paul Santmire; *This Little Planet* edited by M. Hamilton; and *The Doomsday Book* by G. R. Taylor.
(Topics):popular.

Bushwick, N. (1985, June). Land and conservation. *Engage/Social Action*, pp. 26-33.
(Topics):land, conservation.

Bye, B. (1977). *What about lifestyle?* Exeter, England: Paternoster Press.
(Topics):lifestyle.

Byron, W. J. (1975). *Toward stewardship: an interim ethic of poverty, pollution and power*. New York: Paulist Press.
(Topics):stewardship, pollution.

Byron, W. J. (1978). The ethics of stewardship. In M. E. Jegen & B. V. Manno (Eds.), *The Earth is the Lord's: essays on stewardship* (pp. 44-50). New York: Paulist Press.
(Topics):ethics, stewardship.

Cable, T. (1987). Environmental education at Christian colleges. *Perspectives on Science and Christian Faith*, 39(3), 165-168.
(Topics):popular.

Calhoun, R. (1972). *Ecology and Christian responsibility: a study of the theological and ethical bases for Christian ecological concern*. Th.D., Southwestern Baptist Theological Seminary.
(Topics):theology, ethics.

Callicott, J. B. (1990). Genesis and John Muir. *ReVision*, 12(Winter):31-46.
(Topics):ethics.

Campbell, J. B. (1990). Our common future. *Church and Society*, 81(Sept./Oct.), 31-40.
(Topics):popular.

Campbell, N. (1972). Start where you can. *Frontier*, 15(8), 154.
(Topics):popular.

Canadian Scientific and Christian Affiliation (Guelph Chapter) (1984). Christian guidelines for biotechnology. *Journal of the American Scientific Affiliation*, 36(1), 39.
(Topics):genetic engineering.

Carmody, J. (1980). *Theology for the 1980's*. Philadelphia: Westminister Press.
(Topics):theology.

Carmody, J. (1983). *Ecology and religion: toward a new Christian theology of nature*. New York: Paulist Press.
(Topics):theology, book.

Carney, V. (1976). Response to James Donaldson. *Foundations: A Baptist Journal of History and Theology*, 19(3), 252-253.
(Topics):popular.

Carothers, J. E., M. Mead, D. D. McCracken, & R. L. Shinn (Eds.) (1972). *To love or to perish: the technological crisis of the churches*. New York: Friendship Press, Inc.
(Topics):technology, book.

Carpenter, D. K. (1974). Go slow. *Journal of the American Scientific Affiliation*, 26(1), 16.
(Topics):population.

Carpenter, E. (1980). *Animals and ethics*. London: Watkins.
(Topics):animal rights.

Carrick, I. (1970). Right involvement with nature. *Frontier*, 13, 31-33.
(Topics):popular.

Carson, R. (1962). *Silent spring*. Boston: Houghton Mifflin.
Secular work which marks the beginning of the environmental era.
(Topics):popular.

Carter, N. C. (1990). A mother/sister earth pilgrimage. *Daughters of Sarah*, 16(3), 14-17.
(Topics):ecofeminism.

Case, R. (1980). Human responsibility and human liberation. *Journal of the American Scientific Affiliation*, 32(2), 79-83.
(Topics):nuclear.

Cassel, J. F. (1971a). The Christian's role in the problems of contemporary human ecology. In D. R. Scoby (Ed.), *Environmental ethics* (pp. 154-166). Minneapolis: Burgess Publishing Co.
(Topics):ethics.

Cassel, J. F. (1971b). Ecology, God and me. In D. R. Scoby (Ed.), *Environmental Ethics* (pp. 225-227). Minneapolis: Burgess Publishing Co.
(Topics):ethics.

Castro, E.(ed.) (1990). Come Holy Spirit -- renew the whole creation; Giver of life -- sustain your creation. *The Ecumenical Review*, 42(April), 89-174.
Thematic issue for Earthday 1990.
(Topics):popular.

Catherwood, F. (1983). The new technology: the human debate. In J. R. W. Stott (Ed.), *The year 2000*. Downers Grove, IL.: Inter-Varsity Press.
(Topics):technology.

Catholic Bishops (Philippines) (1988a). Think globally, act locally, be spiritually. *The Amicus Journal: A Publication of the Natural Resources Defense Council*, 10(4), 8-10.
(Topics):popular.

Catholic Bishops (Philippines) (1988b, April 15). What is happening to our beautiful land? *SEDOS Bulletin*, pp. 112-115.
(Topics):popular.

Cauthen, K. (1969). The case for Christian biopolitics. *The Christian Century*, 86(47), 1481-1483.
(Topics):popular.

Cauthen, K. (1971). *Christian biopolitics: a credo and strategy for the future.* Nashville: Abingdon Press.
(Topics):book.

Cauthen, K. (1973). Ecojustice: a future-oriented strategy of ministry and mission. *Foundations*, 16, 156-170.
(Topics):popular, ecojustice.

Cauthen, K. (1985a). Imaging the future: new visions and new responsibilities. *Zygon*, 20(3), 321-339.
(Topics):theology.

Cauthen, K. (1985b). Process theology and eco-justice. In D. Hessel (Ed.), *For Creation's sake: preaching, ecology and justice* (pp. 84-85). Philadelphia: Geneva Press.
(Topics):theology, ecojustice.

Cawley, M. (1985). A monastic experience and theology of waste material recycling. *Cistercian Studies*, 20, 237-248.
(Topics):popular, recycling, resources.

Cesaretti, C. A., & Commins, S. (Eds.) (1981). *Let the earth bless the Lord: a Christian perspective on land use.* New York: The Seabury Press.
(Topics):land stewardship, popular book.

Chamberland, D. (1986). Genetic engineering: promise & threat. *Christianity Today*, 30(2), 22-28.
(Topics):genetic engineering.

Chandler, D. H. (1982). Energy: toward more ethical alternatives. *Christian Scholar's Review*, 11(2), 112-123.
(Topics):energy.

Christian Century (The) (1970). The struggle for an ecological theology. *The Christian Century*, 87(9), 275-277.
Special focus issue on the stewardship of Earth.
(Topics):popular.

Christian Ecology Group (1983). *God's green world (a symposium)*. The Christian Ecology Group, 58 Quest Hills Road, Malvern, Worcs. WR14 1RW United Kingdom.
(Topics):Theology.

Christian Medical Society (The) (1970). A Protestant affirmation on the control of human reproduction. *Journal of the American Scientific Affiliation*, 22(2), 46-49.
(Topics):population.

Christianity Today (editorial) (1970a). Ecologism: a new paganism. *Christianity Today*, 14(14), 637-638.
(Topics):popular.

Christianity Today (editorial) (1970b). Fulfilling God's cultural mandate. *Christianity Today*, 14(11), 488-489.
(Topics):popular.

Christianity Today (editorial) (1971). Terracide. *Christianity Today*, 15(15), 706-707.
(Topics):popular.

Christianity Today (interview). (1988) Protecting the Lord's canvas. *Christianity Today*, 32(17), 74, 76.
Interview of Calvin B. DeWitt by Randy Frame.
(Topics):popular.

Christianity Today (anonymous news piece). (1991). Environment: religious leaders join scientists in ecological concerns. *Christianity Today*, 35(9), 49.
On the June 2-3, 1991, meeting of scientists and religious leaders held in New York to continue discussion of a joint effort to address the ecological crisis.
(Topics):popular.

Christiansen, A. (1990). Notes on moral theology, 1989: ecology, justice, and development. *Theological Studies*, 51(1):64-81.
(Topics):ecojustice, economics, theology.

Church, F. F. (1983). Beyond nemesis. *Religious Humanism*, 17, 134-138.
"We are responsible no longer only for our own salvation, but for the cultivation and preservation of life itself. It is not so much that we have arrogated God's prerogative; rather we have inherited certain essential responsibilities once thought to be exclusively a part of the heavenly domain...."
(Topics):popular.

Church of England (Board of Social Responsibility) (1970). *Man in his living environment: an ethical assessment.* Westminster: Church House Publishing.
(Topics):ethics, book.

Church of England (report of the General Synod) (1986). *Our responsibility for the living environment.* London: Church House Publishing.
(Topics):ethics, book.

Church of Scotland (Society, Religion and Technology Project) (1986). *While Earth endures.* SRT Project/Quorum Press.
(Topics):ethics.

Cizik, R. (1990). D.C. dateline: concern & caution. *United Evangelical Action*, 49(May), 17.
(Topics): popular.

Clampit, M. K. (1972). The ecological covenant and nature as a sacred symbol. *Religious Education*, 67(3), 194-198.
(Topics):theology, education.

Clapp, R. (1987). The fate of the soil. *Christianity Today*, 31(14), 14-15.

(Topics):land, popular.

Clark, K. C. (1977) An Analysis of Isaiah 40-44:23 Utilizing the Creation↔Redemption Model of the Creator-King and Process Theology. D. Min., School of Theology at Claremont.
(Topics):theology.

Clark, S. R. L. (1986). Christian responsibility for the environment. *The Modern Churchman*, 28(2), 24-31.
(Topics):popular.

Clark, W. N. (1968). Technology and man: a Christian vision. In I. G. Barbour (Ed.), *Science and religion: new perspectives on the dialogue.* New York: Harper & Row, Publishers.
Utilitarian. Technology is seen as a tool which permits humanity to dominate nature.
(Topics):technology.

Clarke, A. (1989). Urban-rural gardens. *The Egg: A Journal of Eco-Justice*, 9(1), 10-11.
(Topics):agriculture, land.

Clarke, L. R. (1985). The Quaker background of William Bartram's view of nature. *Journal of the History of Ideas*, 46(2), 435-448.
Early American Quaker views of nature.
(Topics):history.

Clifford, R. J. (1985). The Hebrew scriptures and the theology of creation. *Theological Studies*, 46, 507-523.
(Topics):Judaism, theology.

Clobus, R. (1991). Ecofarming and land ownership in Ghana. In G. T. Prance & C. B. DeWitt (Eds.), *Missionary Earthkeeping.* Macon, Georgia: Mercer University Press.
(Topics):popular, land, agriculture, farming.

Cobb, J. B. (1965). *A Christian natural theology*. Philadelphia: Westminster Press.
(Topics):theology.

Cobb, J. B. (1966). *A Christian natural theology*. London: Lutterworth Press.
(Topics):theology.

Cobb, J. B. (1970a). Ecological disaster and the church. *Christian Century*, 87, 1185-1187.
(Topics):popular.

Cobb, J. B. (1970b, Sept. 12). The population explosion and the rights of the subhuman world. *IDOC*, pp. 41-62.
(Topics):population.

Cobb, J. B. (1971). Christian theism and the ecological crisis. *Religious Education*, 66(1), 23-30.
(Topics):popular.

Cobb, J. B. (1972). *Is it too late? a theology of ecology*. Beverly Hills, CA: Bruce.
(Topics):theology, book.

Cobb, J. B. (1973). Ecology, ethics and theology. In H. E. Daly (Ed.), *Toward a steady state economy*. San Francisco: W. H. Freeman and Company.
(Topics):theology, ethics.

Cobb, J. B. (1974a). The Christian, the future, and Paolo Soleri. *The Christian Century*, 91(37), 1008-1011.
    Uses cities designed and built by Paolo Soleri to illustrate that it is not necessary to "sacrifice our relation to nature for the sake of urban values."
(Topics):popular.

Cobb, J. B. (1974b). The local church and the environmental crisis. *Foundations*, 17(2), 164-172.
(Topics):popular, lifestyle.

Cobb, J. B., Jr. (1979). Christian existence in a world of limits. *Environmental Ethics*, 1(2), 149-158.
(Topics):ethics.

Cobb, J. B., Jr. (1980). Process theology and environmental issues. *The Journal of Religion*, 60(4), 440-458.
(Topics):theology.

Cobb, J. B., Jr. (1982). *Process theology as political theology.* Philadelphia: Westminster Press.
(Topics):theology.

Cobb, J. B., Jr. (1984). Envisioning a just and peaceful world. *Religious Education*, 79(4), 483-494.
(Topics):popular.

Cobb, J. B., Jr. (1988). A Christian view of biodiversity. In E. O. Wilson (Ed.), *Biodiversity* (pp. 481-485). Washington, D.C.: National Academy Press.
(Topics):biodiversity.

Cobb, J. B., Jr., & Griffin, D. R. (1976). *Process theology: an introductory exposition.* Philadelphia: Westminster Press.
(Topics):theology.

Cobourn, C. (1990). The struggle to save the Philippines [damaged ecosystem]. *The Witness*, 73(October), 6-9.
(Topics):popular.

Cochrane, C. C. (1984). *The gospel according to Genesis: a guide to understanding Genesis 1-11.* Grand Rapids, MI: Eerdmans.
See p. 28 for comments regarding care for the creation.
(Topics):book.

Cohen, B. L. (1980). Far greater dangers than nuclear. *Journal of the American Scientific Affiliation*, 32(2), 89-92.
(Topics):nuclear.

Cohen, J. (1985). The Bible, man and nature in the history of western thought: a call for reassessment. *The Journal of Religion*, 65(2), 155-172.
(Topics):history.

Cohen, J. (1990). Ecology and social meaning: on classical Judaism and Environmental crisis. *Tikkun*, 5(2), 74-77.
(Topics):Judaism.

Coleman, W. (1976). Providence, capitalism, and environmental degradation: English apologetics in an era of economic revolution. *Journal of the History of Ideas*, 37, 27-44.
(Topics):economics, history.

Compton, E. (1978/79). A plea for a theology of nature. *The Modern Churchman*, ns, 22(1), 3-11.
(Topics):theology, popular.

Compton, T. L. (1982). Natural resources stewardship: the earth is the Lord's. In E. R. Squiers (Ed.), *The environmental crisis: the ethical dilemma* (pp. 109-114). Mancelona, MI: Au Sable Institute of Environmental Studies.
(Topics):resources.

Conrad, A. C. (1954). *The divine economy: a study in stewardship*. Grand Rapids, MI: Eerdmans.
(Topics):theology, stewardship.

Conroy, D. B. (1987a). The Christian family: foundation for environmental renewal: part one: family and community. In F. W. Krueger (Ed.), *Christian ecology: building an environmental ethic for the twenty-first century* (pp. 87-88). North Webster, Indiana: The North American Conference on Christianity and Ecology.
(Topics):popular.

Conroy, D. B. (1987b). Regenerative agriculture: an integrated approach to farming and life. In F. W. Krueger (Ed.), *Christian ecology: building an environmental ethic for the twenty-first century* (p. 37). North Webster, Indiana:

The North American Conference on Christianity and Ecology.
(Topics):land, agriculture.

Conroy, D. B. (1990, February 17). The Church awakens to the global environmental crisis. *America*, pp. 149-152.
Environmentalists have been distressed by the absence of church leaders in their battle to save nature. Many are now openly stating that they want the religious community to get involved.
(Topics):popular.

Cook, H. (1988). Laboratory animals in agriculture. *Earthkeeping: A Quarterly on Faith and Agriculture*, 4(1), 13-14.
(Topics):animal rights.

Cook, H. (1989). GEMs in the strawberry patch. *Earthkeeping: A Quarterly on Faith and Agriculture*, 5(3), 20-21.
(Topics):genetic engineering.

Cooper, R. (1990). Through the soles of my feet: a personal view of creation. *International Review of Mission*, 79(314), 179-185.
(Topics):popular.

Cooper, T. (1990). *Green Christianity: caring for the whole creation*. London: Hodder and Stoughton.
Cooper is an economist and consultant on ecology. He has been co-chair of the Green Party Council (Britain) and twice a parliamentary candidate.
(Topics):popular book, economics.

Cooper, T., & Kemball-Cook, D. (Eds.) (1983). *God's green world*. Christian Ecology Group, out of print.
For more information about the Christian Ecology Group, please write to the Secretary, Mrs. Judith Pritchard, 58 Quest Hills Road, Malvern, Worcs. WR14 1RW.
(Topics):popular.

Cormack, P. (1989, January - March). What is the land saying to you? *Together, A Publication of World Vision*, p. 13.
(Topics):land.

Cowap, C. (1986). Energy production and use: ethical implications. *Engage / Social Action*, 14(4), 11-16.
(Topics):ethics, energy.

Cox, H. (1965). *The secular city*. New York: Macmillan.
Cox is quite utilitarian and fails to emphasize the stewardly responsibility of humanity for creation.
(Topics):popular.

Cox, H. (1966). The responsibility of the Christian in a world of technology. In D. Munby (Ed.), *Economic growth in world perspective*. New York: Association Press.
Technology is seen as a means to master nature.
(Topics):popular, technology.

Cox, K. M. (1979). *A comparison of the Biblical and Native American views of the human relationship with nature*. Ph.D., Graduate Theological Union.
(Topics):theology.

Crabtree, A. (1977). To be or not to be. *Journal of Ecumenical Studies*, 14(1), 92-96.
(Topics):popular.

Crabtree, A. B. (1972). The urgency of ecological ecumenism. *Journal of Ecumenical Studies*, 9, 94-97.
(Topics):popular.

Crabtree, A. B. (1973). Ecological crisis and cultural change. *Journal of Ecumenical Studies*, 10, 576-580.
(Topics):popular.

Craine, R. (1987). Hildegard of Bingen: a sign for our times. In F. W. Krueger (Ed.), *Christian ecology: building an environmental ethic for the twenty-first century* (pp. 25-

26). North Webster, Indiana: The North American Conference on Christianity and Ecology.
(Topics):popular.

Creator's Mandate (The), 150-207 Castlegreen Dr., Thunder Bay, Ont. P7A 7L8, Canada; (807) 767-9237.
A bi-monthly newsletter for evangelicals focusing on issues of the environment.
(Topics):popular.

Crotty, N. (1971). Catholic moral theology and ecological responsibility. *Religious Education*, 66(1), 44-49.
(Topics):popular.

Crowley, M. (1985). The virtues: commitment, spiritual practice and transformation. *Epiphany*, 6(1), 52-59.
(Topics):popular.

Cubbage, R. (1988, October). Miriam Therese MacGillis: to love and save the earth. *St. Anthony Messenger*, pp. 28-35.
(Topics):popular.

Cumbey, C. (1983). *The hidden dangers of the rainbow: the New Age Movement and our coming age of barbarism.* Shreveport, LA: Huntington House.
Fundamentalist author who is highly critical of the awakening of environmental responsibility within the church and tries to link the movement and many of its leaders with the non-Christian "New Age" movement.
(Topics):book.

Cummings, C. (1989). Exploring eco-spirituality. *Spirituality Today*, 41, 30-41.
(Topics):popular.

Cundy, I. (1977). The church as community. In I. Cundy (Ed.), *Obeying Christ in a changing world, Vol. 2, the people of God.* Glasgow: Collins.
(Topics):popular.

Cupitt, D. (1990). The value of life. *The Modern Churchman*, 32(2), 39-45.
(Topics):ethics

Curley, E. (1972, February). A short theology of ecology. *Religion Teacher's Journal*, pp. 18-19.
(Topics):popular.

Curry-Roper, J. M. (1990). Contemporary Christian eschatologies and their relation to environmental stewardship. *The Professional Geographer*, p. 161.
(Topics):worldview, stewardship.

Daetz, D. (1970a). Heading off eco-catastrophe: a job for new Adam. *Lutheran Quarterly*, 22, 271-278.
(Topics):popular.

Daetz, D. (1970b). No more business as usual. *Dialog*, 9(3), 171-175.
(Topics):popular.

Dahl, N. A. (1952). The parables of growth. *Studia Theologica*, 5, 132-166.
(Topics):theology.

Dalai Lama, the (1987). An ethical approach to environmental protection. In S. Davies (Ed.), *Tree of life: Buddhism and protection of nature with a declaration on environmental ethics from his holiness the Dalai Lama and an introduction by Sir Peter Scott* (p. 5). Buddhist Perception of Nature.
(Topics):Buddhism.

Dalton, M. A. (1976). Theology of ecology: an interdisciplinary concept. *Religious Education*, 71, 17-26.
(Topics):theology, popular.

Dalton, A. M. (1990). Befriending an estranged home.

Religious Education, 85(1), 15-24.
(Topics):education.

Daly, H. E. (1973). *Toward a steady state economy*. San Francisco: W. H. Freeman and Company.
Major contribution to an economic system designed for global sustainability. A secular work but undergirded by assuming the validity of Biblical principles.
(Topics):economics.

Daly, H. E. (1977). *Steady-state economics*. San Francisco: W. H. Freeman and Co.
An economic system compatible with sustainability and undergirded by Biblical principles on the nature of man, justice, and creation.
(Topics):economics.

Daly, H. E. (1980a). The ecological and moral necessity of limiting growth. In R. L. Shinn (Ed.), *Faith and science in an unjust world. Report of the World Council of Churches' conference on faith, science and the future, Vol. 1: plenary presentations*. Philadelphia: Fortress Press.
(Topics):economics.

Daly, H. E. (Ed.) (1980b). *Economics, ecology, ethics: essays towards a steady-state economy*. San Francisco: W. H. Freeman & Co.
Twenty-two essays on the ethical and physical principles of steady-state economics. Contributors include Kenneth Boulding, Garrett Hardin, Paul and Anne Ehrlich, E. F. Schumacher, and Nichols Georgescu-Roegen. An important secular work on a sustainable economic system.
(Topics):economics.

Daly, H. E., & Cobb, J. B., Jr. (1989). *For the common good: redirecting the economy toward community, the environment, and a sustainable future*. Boston: Beacon Press.
Herman Daly, economist for the World Bank, is a specialist in steady-state economic theory. John Cobb, Jr., is a theologian. The topic of their book is the development of an

economic system that will work in conjunction with and assist in forging a sustainable future.
(Topics):economics.

Damaskinos, Met. of Switzerland (1990). Man and his environment: The ecological problem: its positive and negative aspects [two articles]. *The Journal of the Moscow Patriarchate*, 85(Summer), 412-423.
(Topics):popular.

Dammers, H. (1986). *A Christian life style.* London: Hodder and Stoughton.
(Topics):lifestyle.

Dannacker, E. (1988). Airborne connection to Brazil. *Earthkeeping: A Quarterly on Faith and Agriculture*, 4(1), 20-21.
Discusses the impact of deforestation in the Amazon basin on the purple martin, a winter resident.
(Topics):biodiversity.

Dannen, K. (1972). Wilderness and the reformers. *Lexington Theological Quarterly*, 7(4), 103-112.
(Topics):theology, wilderness.

Dantine, W. (1965). Creation and redemption: an attempt at a theological understanding in light of contemporary understanding of the world. *Scottish Journal of Theology*, 18, 129-147.
(Topics):theology.

Davies, S. (ed.) (1987). *Tree of life: Buddhism and protection of nature with a declaration on environmental ethics from his holiness the Dalai Lama and an introduction by Sir Peter Scott.* Buddhist Perception of Nature.
(Topics):Buddhism.

Davies, W. D. (1974). *The gospel and the land: early Christianity and Jewish territorial doctrine.* Berkeley, Calif.: University of California Press.
(Topics):land, theology.

Davies, W. D. (1982). *The territorial dimension of Judaism.* Berkeley: University of California Press.
(Topics):land, theology, Judaism.

Davis, D. J. (1974). A selected bibliography of the writings of Joseph Sittler. *The Journal of Religion,* 54, 181-183.
(Topics):theology.

Davis, M. (1987). War, conflict and environmental degradation: part three: the effects of militarism. In F. W. Krueger (Ed.), *Christian ecology: building an environmental ethic for the twenty-first century* (p. 76). North Webster, Indiana: The North American Conference on Christianity and Ecology.
(Topics):nuclear.

Davis, W. H. (1972). The ecological crisis. *Review and Expositor,* 69(1), 5-9.
(Topics):popular.

Daw, R. W. (ed.) (1986). Economic justice for all: Catholic social teaching and the U. S. economy, third draft. *Origins,* 16(3), 33-76.
(Topics):popular, economics.

De la Cruz, A. (1988). Scriptural basis of ecology: a mandate for environmental stewardship. *Taiwan Journal of Theology,* 10, 211-223.
(Topics):theology.

de Roo, R. (1986). Agri-culture, agri-business, agri-power: a look to the future. *Earthkeeping: A Quarterly on Faith and Agriculture,* 2(4), 4-8.
(Topics):agriculture, land.

Deloria, V. (1973). *God Is Red*. New York: Grossett.
Suggests that Native American religions are more appropriate than Christianity in addressing American social and environmental issues.
(Topics):popular.

DeMott, J. (1990). Make environment ministry a priority. *The Witness*, 73(September), 20-21.
(Topics):popular.

Denig, N. W. (1985). "On values" revisited: a Judeo-Christian theology of man and nature. *Landscape Journal*, 4(2), 96-105.
(Topics):theology.

Denisenko, F., Met. of Kiev and Galich (1985). Global threat to mankind -- global strategy of peace. *The Journal of the Moscow Patriarchate*, 10, 29-40.
(Topics):popular, nuclear, pollution.

Derr, T. S. (1970). Man against nature. *Cross Currents*, 20, 263-275.
(Topics):theology.

Derr, T. S. (1973). *Ecology and human liberation: a theological critique of the use and abuse of our birthright*. Vol. III, no.1, serial no.7. Geneva: World Student Christian Federation (WSCF).
Cooperative publication by WSCF and the World Council of Churches.
(Topics):book.

Derr, T. S. (1975a). *Ecology and human need*. Philadelphia: Westminster.
Revised version of Derr 1973.
(Topics):book.

Derr, T. S. (1975b). Religion's responsibility for the ecological crisis: an argument run amok. *Worldview*, 18(1), 39-45.
(Topics):popular.

Derr, T. S. (1977). Conversations about ultimate matters: theological motifs in ecumenical studies of the technological future. *International Review of Missions*, 66, 123-134.
(Topics):technology.

Derr, T. S. (1981). The obligation to the future. In E. Partridge (Ed.), *Responsibilities to future generations: environmental ethics*. Buffalo, N.Y.: Prometheus Books.
(Topics):ethics.

Derr, T. S. (1983). *Barriers to ecumenism: the Holy See and the World Council of Churches on social questions*. Maryknoll, N.Y.: Orbis Books.
(Topics):book.

Derrick, C. (1972). *The delicate creation: towards a theology of the environment*. Old Greenwich, Conn.: Devin-Adair Company.
(Topics):book, theology.

Devadas, D. (1986). Work, technology and the environment. *One World*, pp. 11-15.
(Topics):technology.

De Vries, G., Jr. (1989). The heavens again. *The Reformed Journal*, 39(10), 9.
Expresses concern for the lack of passion for environmental matters among the rank-and-file Christians and the majority of the pastorate.
(Topics):popular.

Dew, W. H. (1950). Religious approach to nature. *Church Quarterly Review*, 150, 81-99.
An important early work addressing the lack of a proper biblically based attitude toward nature in western Christianity.
(Topics):theology.

DeWitt, C. B. (1982). Stewardship and justice in the exploitation of deep seabed minerals. In E. R. Squiers (Ed.), *The environmental crisis: the ethical dilemma* (pp. 151-176). Mancelona, MI: Au Sable Institute of Environmental Studies.
(Topics):resources.

DeWitt, C. B. (1986, February 26). Earthkeeping and the kingdom of God. *The Banner.*
(Topics):popular.

DeWitt, C. B. (1987a). A Christian land ethic. *The Banner,* 122(43), 10.
(Topics):land.

DeWitt, C. B. (1987b). Christian stewardship: its basis in the cosmos, the scriptures and spirituality. In F. W. Krueger (Ed.), *Christian ecology: building an environmental ethic for the twenty-first century* (pp. 20-21). North Webster, Indiana: The North American Conference on Christianity and Ecology.
(Topics):popular.

DeWitt, C. B. (1987c). Responding to creation's degradation: scientific, scriptural and spiritual foundations. In *Proceedings of the NACCE.* San Francisco: N.A. Conf. on Christianity & Ecology.
(Topics):popular.

DeWitt, C. B. (1987d). Statement by Dr. Calvin DeWitt at press conference. In F. W. Krueger (Ed.), *Christian ecology: building an environmental ethic for the twenty-first century* (p. 6). North Webster, Indiana: The North American Conference on Christianity and Ecology.
(Topics):popular.

DeWitt, C. B. (1987e). *A sustainable earth: religion and ecology in the western hemisphere.* Au Sable Institute of Environmental Studies, Mancelona, MI.
Available through Au Sable Institute.

(Topics):popular.

DeWitt, C. B. (1988a). *The Sabbath's meaning.* New York: United Nations.
A preface to the 1988 folder for the Environmental Sabbath, United Nations Environmental Programme.
(Topics):popular.

DeWitt, C. B. (1988b). Why a Christian ecology. *Firmament: The Quarterly of Christian Ecology,* 1(1), 2-3.
(Topics):popular.

DeWitt, C. B. (1988c). You will live faithfully in the land. *Epiphany,* 9(1), 60-68.
(Topics):popular.

DeWitt, C. B. (1989a). Christian stewardship. *The Trumpeter,* 6(4), 170-171.
(Topics):popular.

DeWitt, C. B. (1989b). Ecological issues and our spiritual roots. *Christian Living,* 36(10), 14-18.
(Topics):popular.

DeWitt, C. B. (1989c, Fall). The price of gopher wood. *Faculty Dialogue,* pp. 59-62.
(Topics):Noah, popular, biodiversity.

DeWitt, C. B. (1989d). Seven degradations of creation. *Perspectives: A Journal of Reformed Thought,* 4(2), 4-8.
Paper given at Madison, Wisconsin, conference sponsored by NACCE and World Alliance of Reformed Churches -- Oct. 11-15, 1988.
(Topics):ozone, greenhouse effect, land, water, habitat destruction, species extinction, wastes, culture, popular, biodiversity, pollution.

DeWitt, C. B. (1990a). Assaulting the gallery of God: humanity's seven degradations of the earth. *Sojourners,* 19(2), 19-21.

(Topics):ozone, greenhouse effect, land, water, habitat destruction, species extinction, wastes, culture, popular, biodiversity, pollution.

DeWitt, C. B. (1990b, April). Can we help save God's Earth? *ESA Advocate*, pp. 12-13.
(Topics):popular.

DeWitt, C. B. (1990c). A crisis in Creation. *The Church Herald*, 47(3), 8-10.
(Topics):popular.

DeWitt, C. B. (1990d). Seven degradations of creation: challenging the church to renew the covenant. *Firmament*, 2(1), 5-9.
(Topics):ozone, greenhouse effect, land, water, habitat destruction, species extinction, wastes, culture, popular, biodiversity, pollution.

DeWitt, C. B. (Ed.) (1991a). *The environment and the Christian: what does the New Testament say about the environment?* Grand Rapids, Michigan: Baker Book House.
An important major theological work.
(Topics):theology.

DeWitt, C. B. (1991b). The religious foundations of ecology. In J. Scherff (Ed.), *The mother earth handbook*. New York: Continuum Publishing Company, pp. 248-268.
Reviews the major world religions' positions on the environment with a special emphasis on the Hebrew Bible and New Testament perspectives.
(Topics):theology, popular.

DeWitt, C. B. (1991c). The Church's role in environmental action. *Word & World*, 11(2):180-185.
(Topics):theology, popular.

DeWolf, L. H. (1970). Christian faith and our natural environment. *IDOC International*, 9, 3-15.

(Topics):popular, pollution, population.

DeWolf, L. H. (1971a). Christian education for survival on Earth. *Religious Education*, 66(1), 23-30.
(Topics):popular, theology.

DeWolf, L. H. (1971b). *Responsible freedom: guidelines to Christian action*. New York: Harper and Row.
See especially chapter 11.
(Topics):popular.

DeWolf, L. H. (1971c). Theology and ecology. *American Ecclesiastical Review*, 164, 154-170.
(Topics):theology.

Dickenson, I. (1987). Appropriate ecological programs for pastors and churches. In F. W. Krueger (Ed.), *Christian ecology: building an environmental ethic for the twenty-first century* (p. 101). North Webster, Indiana: The North American Conference on Christianity and Ecology.
(Topics):popular, education.

Diener, R. (Ed.) (1970). *The Earth day year: religion and ecology, 1970 -- A bibliography.* Ronald Diener, Librarian of the Boston Theological Institute, 45 Francis Avenue, Cambridge, Massachusetts 02138.
The publication date is in question. I have not been able to locate this.
(Topics):theology, popular.

Diers, J. (1990). Humanity and environment: precedence. *Christianity and Crisis*, 50(7):146-147.
Report on the World Council of Churches assembly at Seoul, Korea, in 1990 on the topic of JPIC (Justice, Peace, and the Integrity of Creation).
(Topics):popular.

Dingman, M. J. (1986). Erosion in the vineyard: a call to preserve the family farm. *Sojourners*, 15(9), 26-27.
(Topics):agriculture, land.

Ditmanson, H. H. (1964). The call for a theology of creation. *Dialog*, 3, 264-273.
(Topics):theology, history.

Doan, G. E. (1971). Toward a life-style environmentally informed. *Lutheran Quarterly*, 23, 305-316.
(Topics):lifestyle.

Dobel, J. P. (1977). Stewards of the earth's resources: a Christian response to ecology. *The Christian Century*, 94(32), 906-909.
(Topics):resources, popular.

Dobson, E. G. (1986) *An analysis of the environmental perceptions of undergraduate students in evangelical and fundamentalist Bible colleges and liberal arts colleges.* Ed.D., University of Virginia.
(Topics):knowledge, attitudes, education.

Dodd, C. H. (1946a). Natural law in the Bible -- I. *Theology*, 49(311), 130-133.
(Topics):theology.

Dodd, C. H. (1946b). Natural law in the Bible -- II. *Theology*, 49(312), 161-167.
(Topics):theology.

Donahue, J. R. (1977). Biblical perspectives on justice. In J. C. Haughey (Ed.), *In the faith that does justice.* New York: Paulist Press.
(Topics):book, ecojustice.

Donaldson, J. (1976). America the beautiful: interdependence with nature (with replies by V. Carney and G. D. Hammer). *Foundations*, 19, 238-256.
(Topics):popular.

Dorr, D. (1984). *Spirituality and justice.* Maryknoll, NY: Orbis Books.

Brief discussion of lifestyle and ecological issues as part of the broad theme of justice.
(Topics):lifestyle, ecojustice.

Dorr, D. (1987). To care for the Earth. *Furrow*, 38, 454-460.
Extensive review of the book *To care for the Earth: a call to a new theology* by Sean McDonagh.
(Topics):land.

Douglas, D. (1981). God, the world and James Watt. *Christianity and Crisis: A Christian Journal of Opinion*, 41, 258+.
(Topics):popular.

Douglas, D. (1984). Wild country and wildlife: a spiritual preserve. *Christian Century*, 10(1), 11-13.
Discusses the importance and value of wilderness.
(Topics):popular, wilderness.

Dowd, M. (1990). *The meaning of life in the 1990's: an ecological, Christian perspective.* Granville, MA: Living Earth Christian Fellowship.
Call (413)357-6111 for information.
(Topics):popular.

Doyle, B. (1983). *Meditations with Julian of Norwich.* Santa Fe: Bear and Company.
(Topics):creation spirituality.

Dubos, R. (1972). *A god within.* New York: Charles Scribner's.
(Topics):theology, book.

Dubos, R. (1973a). Francis vs. Benedict: the first was a conservationist, the second a good steward. *Catholic Digest*, 37, 38-43.
Excerpts from *A God Within.*
(Topics):popular, St. Francis of Assisi, St. Benedict of Nursia.

Dubos, R. (1973b). A theology of Earth. In I. G. Barbour (Ed.), *Western man and environmental ethics* (p. 275). Reading, Mass.: Addison-Wesley Publishing Company.
(Topics):theology.

Dubos, R. (1980). *The wooing of Earth.* New York: Charles Scribner's Sons.
(Topics):popular book.

Duchrow, U., and Liedke, G. (1989). *Shalom: biblical perspectives on creation, justice & peace.* Geneva: WCC Publications.
(Topics):ecojustice, popular book.

Dumas, A. (1975). The ecological crisis and the doctrine of creation. *The Ecumenical Review,* 27, 24-35.
(Topics):theology, resources, pollution, ethics.

Dumas, A. (1977). The creation--God's glory in His world. *The Reformed World,* 34(7-8), 311-315.
(Topics):popular.

Dumas, A. (1978). A society which creates justice: three themes but one development. *The Ecumenical Review,* 30, 211-219.
(Topics):popular, ecojustice.

Dumas, A. (1979). When science looks to faith. *The Ecumenical Review,* 31, 388-393.
(Topics):popular.

Dumbrell, W. J. (1984). *Covenant and creation: a theology of Old Testament covenants.* Nashville, TN: Thomas Nelson Publishers.
(Topics):theology.

Dumbrell, W. J. (1985). Genesis 1-3, ecology, and the dominion of man. *Crux,* 21(4), 16-26.
(Topics):theology.

Duncan, R. (1976). Adam and the ark. *Encounter*, 37, 189-197.
    (Topics):biodiversity.

Dunn, S. (1990). Ecology, ethics, and the religious educator. *Religious Education*, 85(1), 34-41.
    (Topics):ethics, creation spirituality.

Duraisingh, C. (1990). The Holy Spirit, mission and renewal of the earth [editorial]. *International Review of Mission*, 79(314), 137-141.
    Addresses the World Council of Church's environmental agenda for the VIIth Assembly in Canberra.
    (Topics):popular.

Dusilek, D. (1987, April-June). Land -- more than plain ground. *Together: A Publication of World Vision*, pp. 1-2.
    (Topics):land.

Dyrness, W. (1987). Stewardship of the Earth in the Old Testament. In W. Granberg-Michaelson (Ed.), *Tending the garden: essays on the Gospel and the Earth* (pp. 50-65). Grand Rapids, MI: Eerdmans Publishing Company.
    (Topics):theology, popular, stewardship.

Dyrness, W. A. (1991). Are we our planet's keeper? *Christianity Today*, 35(4), 40-42.
    (Topics):popular.

Dyson, M. L. (1990). Ecological metaphors in feminist eschatology. *Daughters of Sarah*, 16(3), 24-27.
    (Topics):ecofeminism.

Echlin, E. P. (1973). Ministry to the environment. *Priest*, 29, 18-21.
    (Topics):popular.

Echlin, E. P. (1974). Let the earth rejoice. *Frontier*, 17, 167-170.
    (Topics):popular.

Echlin, E. P. (1989). Ecumenical ecology at Windsor Castle. *Month*, 22, 78-80.
(Topics):popular.

Edinburgh, The Duke of, and Mann, Michael (the Rt. Rev.). (1989). *Survival or extinction: a Christian attitude to the environment.* Windsor Castle, England: St. George's House.
(Topics):popular book.

Ehlers, V. J. (1980). Gems of wisdom and wrong conclusions. *Journal of the American Scientific Affiliation*, 32(2), 78-79.
(Topics):nuclear.

Ehlers, V. J. (1982). Christian stewardship of energy resources -- twenty theses. In E. R. Squiers (Ed.), *The environmental crisis: the ethical dilemma* (pp. 331-344). Mancelona, MI: Au Sable Institute of Environmental Studies.
(Topics):energy.

Ehrenfeld, D. (1978). *The Arrogance of Humanism.* New York: Oxford University Press.
Finds humanism as a world view to be incapable of adequately responding to the environmental crisis.
(Topics):popular book.

Ehrenfeld, D. (1990). Environmental control and the decline of reality. *The Modern Churchman*, 32(2), 5-19.
(Topics):popular.

Ehrenfeld, D., & Bentley, P. J. (1982). Nature in the Jewish tradition: the source of stewardship. *Radix*, 14(3), 10-20.
(Topics):Judaism, stewardship.

Ehrenfeld, D., & Bentley, P. J. (1985). Judaism and the practice of stewardship. *Judaism: A Quarterly Journal of Jewish Life and Thought*, 34, 301-311.
(Topics):Judaism, stewardship, theology.

Elder, F. (1970). *Crisis in Eden: a religious study of man and environment.* Nashville: Abingdon Press.
(Topics):popular book.

Elder, F. M. (1978). Responses to the ecological question: a Christian review. *The Harvard Theological Review,* 71, 318-319.
(Topics):theology.

Elert, W. (1962). *The structure of Lutheranism.* St. Louis: Concordia.
(Topics):theology.

Elliott, R., & Gare, A. (Eds.) (1983). *Environmental philosophy: a collection of readings.* University Park, PA: Pennsylvania State University Press.
(Topics):ethics.

Ellul, J. (1964). *The technological society.* New York: Vintage Books.
(Topics):technology.

Elsdon, R. (1981). *Bent world: a Christian response to the environmental crisis.* Downers Grove, IL: Inter-Varsity Press.
(Topics):book.

Elsdon, R. (1989). A still-bent world: some reflections on current environmental problems. *Science and Christian Belief,* 1(2), 99-121.
(Topics):popular, theology.

Elshtain, J. B. (1990). Nature's call. *First Things,* 7(Nov.), 7-9.
Editorial on Christianity's ecological responsibility.
(Topics):popular.

Elwood, D. J. (1971). Primitivism or technocracy: must we choose? *The Christian Century*, 88(48), 1413-1418.
(Topics):popular, technology.

Engel, D. E. (1970). Elements in a theology of environment. *Zygon*, 5, 216-228.
(Topics):theology.

Engel, J. R. (1987). Teaching the eco-justice ethic: the parable of the Billerica Dam. *The Christian Century*, 104(16), 466-469.
(Topics):ethics, popular.

Engel, J. R. (1988). Ecology and social justice: the search for a public environmental ethic. In W. R. Copeland & R. D. Hatch (Eds.), *Issues of justice: social sources and religious meaning*. Macon, GA: Mercer University Press.
(Topics):ethics.

Engel, J. R., & Bakken, P. *Ecology, justice, and the Christian faith: a guide to the literature 1960-1990*. Chicago: Center for the Scientific Study of Religion.
This forthcoming work is an extensive annotated bibliography on ecojustice.
(Topics):bibliography, ecojustice.

Engel, J. R., & Engel, J. G. (Eds.) (1990). *Ethics of environment and development: global challenge, international response*. Tucson, AZ: The University of Arizona Press.
(Topics):ethics.

Engle, M. C. (1988). Alternative voices: provocative thinking for a theology of stewardship. *Anglican Theological Review*, 70, 91-93.
(Topics):theology.

Ensigns, S. E. (1982). The response of the church to shrinking petroleum availability. In E. R. Squiers (Ed.), *The environmental crisis: the ethical dilemma* (pp. 315-330).

Mancelona, MI: Au Sable Institute of Environmental Studies.
(Topics):energy.

Epiphany (1981). The eleventh commandment. *Epiphany: A Journal of Faith and Insight*, 1(4).
Thematic issue.
(Topics):popular.

Epiphany (1983). Stewardship of the Earth: a Christian response to a spiritual crisis. *Epiphany: A Journal of Faith and Insight*, 3(3), 2-87.
Thematic issue.
(Topics):popular.

Epiphany (1985). To be Christian is to be ecologist. *Epiphany: A Journal of Faith and Insight*, 6(1), 1-89.
Thematic issue.
(Topics):popular.

Epiphany (1988). Christian ecology: a new movement and its first challenge. *Epiphany: A Journal of Faith and Insight*, 8(2), 6-63.
Thematic issue focusing on the first annual conference of the North American Conference on Christianity and Ecology (NACCE).
(Topics):popular.

Episcopal Church (1979). *The book of common prayer and administration of the sacraments and other rites and ceremonies of the church.* New York: Oxford University Press.
Several sections refer to environmental stewardship.
(Topics):popular, Episcopal.

Episcopal Pastoral Letter (1987). Our relationship with nature. *Latin American Documentation*, 18, 12-22.
(Topics):popular.

Erickson, D. and Erickson, D. (1991). *Seven days to care for God's world: Rupert learns what it means to take care of God's good earth.* Minneapolis: Augsburg. Illustrated by Kathy Rogers. Early grade level book on caring for creation. (Topics):education.

Erickson, M. (1978). Human engineering and Christian ethical values. *Journal of the American Scientific Affiliation,* 30(1), 16-20. (Topics):genetic engineering.

Eternity (1970). What are we doing to God's Earth. *Eternity,* 21(5), 12ff. Thematic issue. (Topics):popular.

Eternity [Editorial] (1987). Cover: who will clean up America's deadly leftovers? *Eternity,* 38(7/8), 10-16. (Topics):toxic wastes.

Ette, A., & Waller, R. (1978). The anomaly of a Christian ecology. *Ecologist Quarterly,* 2(Summer), 144-148. (Topics):popular.

Evangelicals for Social Action and Sider, R. J. (1987). *Completely Pro-Life.* Downers Grove, IL: Inter-Varsity. (Topics):pro-life.

Evans, B. F. (1987). Introduction. In B. F. Evans & G. D. Cusack (Eds.), *Theology of the land* (pp. 9-11). Collegeville, MN: The Liturgical Press. (Topics):land.

Evans, B. F., & Cusack, G. D. (Eds.). (1987). *Theology of the Land.* Collegeville, MN: The Liturgical Press. (Topics):land, theology.

Evans, J. (1990). Use and abuse of tropical forests. *Science & Christian Belief,* 2(2), 141-144.

(Topics):resources, forests.

Evdokimov, P. (1965). Nature. *Scottish Journal of Theology*, 18(1), 1-22.
(Topics):theology.

Event (1970). Man and his environment. *Event*.
Thematic issue.
(Topics):popular.

Everett, W. (1979). *Land ethics: toward a covenantal model*.
Newton Centre, Mass.: The American Society of Christian Ethics.
(Topics):ethics, land.

Fackre, G. (1971). Ecology and theology. *Religion in Life*, 40, 210-224.
(Topics):theology.

Fagley, R. (1960). *The population explosion and Christian responsibility*. New York: Oxford University Press.
(Topics):population.

Fagley, R. (1962). The Christian's response to the population explosion. *Journal of the American Scientific Affiliation*, 14(1), 17-24.
(Topics):population.

Falcke, H. (1984). Confronting threats to peace and survival: theological aspects. *The Ecumenical Review*, 36, 33-42.
(Topics):ethics.

Falcke, H. (1986). Biblical aspects of the process of mutual commitment. *The Ecumenical Review*, 38, 257-264.
(Topics):popular.

Falcke, H. (1987). The integrity of creation. *One World*, pp. 15-18.
(Topics):popular.

Fandrich, H. (1973). The engineer, the consumer and pollution. *Journal of the American Scientific Affiliation*, 25(1), 17-20.
(Topics):pollution.

Faramelli, N. J. (1970). Ecological responsibility and economic justice: the perilous links between ecology and poverty. *Andover Newton Quarterly*, 11(2), 81-93.
(Topics):economics, pollution.

Faramelli, N. J. (1971). *Technethics*. New York: Friendship Press.
(Topics):ethics, technology.

Faramelli, N. J. (1972). Ecological responsibility and church investments. *Church and Society*, 62, 25-33.
(Topics):economics, ethics, pollution.

Faramelli, N. J. (1973). The Role of the Church in Eco-justice. *Church and Society*, 64(November/December), 4-15.
(Topics):economics, popular.

Faramelli, N. J. (1974). Ecojustice: challenge to the churches in the 1970's. *Foundations*, 17(2), 124-132.
(Topics):popular, ecojustice.

Faricy, R. (1982). *Wind and sea obey him*. London: SCM Press.
(Topics):book.

Faricy, R. L. (1988). The person-nature split: ecology, women, and human life. *The Teilhard Review*, 23, 33-44.
(Topics):popular.

Farrell, M. (1981). J. Watt: occupy the land until Jesus comes. *National Catholic Reporter*, 17, 16.
(Topics):popular.

Feddema, R. (1991). Sustainable agriculture. *Earthkeeping: a quarterly on faith and agriculture,* 6(3):19.
(Topics):land, agriculture.

Feenstra (1969). The spiritual vs. material heresy. *Journal of the American Scientific Affiliation,* 21(2), 44-46.
(Topics):popular.

Ferkiss, V. C. (1972). Technology and the future of man. *Review and Expositor,* 69(1), 49-54.
(Topics):technology.

Ferkiss, V. C. (1974). Christianity, technology and the human future. *Dialog,* 13, 258-263.
(Topics):technology.

Ferre, F. (1976). *Shaping the future.* New York: Harper & Row.
Based on a "pluralistic religious stance" which he calls Polymythic Organicism.
(Topics):popular book.

Ferre, F. (1982). Religious world modeling and postmodern science. *The Journal of Religion,* 62, 261-271.
(Topics):ethics, popular.

Ferre, F. (1983). Faith for the future. *American Journal of Theology and Philosophy,* 4(1), 3-13.
(Topics):popular.

Filippi, L. J. (1990). *Of sweet grapes, wheat berries and simple meeting: feminist theology, gestalt therapy, pastoral counseling, and the earth (ecology).* Ph.D., School of Theology at Claremont.
(Topics):theology, ecofeminism.

Finger, R. H. (1990). Many strands -- one web [editorial]. *Daughers of Sarah,* 16(May/June), 2.
(Topics), ecofeminism.

Finnerty, A. (1977). *No more plastic Jesus: global justice and Christian lifestyle.* New York: Orbis Books.
(Topics):lifestyle, technology, book.

Finnin, W., & Huisingh, D. (1972). Population control begins with you. *Duke Divinity Review*, 37, 32-39.
(Topics):population.

Fisher, J. (1972). *Scripture animals: a natural history of the living creatures named in the Bible.* New York: Weathervane Books.
(Topics):scriptural plants and animals.

Fisher, L. (1970). Man and nature in Old Testament traditions. *IDOC International*, 9, 16-39.
(Topics):theology.

Fisher, W. E. (1976). *A new climate for stewardship.* Nashville, TN: Abingdon Press.
(Topics):stewardship.

FitzMaurice, E. (1989). A time to heal. *Momentum*, 20, 75-77.
(Topics):popular.

Foltz, B. V. (1984). On Heidegger and the interpretation of environmental crisis. *Environmental Ethics*, 6(4), 323-338.
(Topics):ethics.

Forbes, J. A. (1985). Preaching in the contemporary world. In D. Hessel (Ed.), *For Creation's sake: preaching, ecology and justice* (pp. 45-54). Philadelphia: Geneva Press.
(Topics):theology, ecojustice.

Forrester-Brown, J. A. (1920). *The two creation stories in Genesis: a study of their symbolism.* London: John M. Watkins -- reprinted in 1974 by Shambala, Berkeley, CA.
(Topics):theology.

Foster, R. J. (1978). *Celebration of discipline.* New York: Harper & Row.

(Topics):lifestyle.

Foster, R. J. (1981). *Freedom of simplicity.* New York: Harper & Row.
(Topics):lifestyle.

Foster, W. J., & Jeays, D. R. (1981). Environmental attitudes, religious beliefs and denominational affiliation in a group of primary teacher education students. *Journal of Christian Education,* 72, 46-51.
(Topics):environmental attitudes, education.

Foundations (1974). Ecology and Justice. *Foundations,* 17, 99-172.
Thematic issue.
(Topics):popular, ethics, ecojustice.

Fox, M. (1979). *Western spirituality: historical roots, ecumenical roots.* Notre Dame, IN: Fides/Claretian.
(Topics):creation spirituality.

Fox, M. (1980). *Breakthrough: Meister Eckhart's creation spirituality in new translation.* Garden City, N.Y.: Doubleday, Image.
(Topics):creation spirituality.

Fox, M. (1983a). *Meditations with Meister Eckhart.* Santa Fe: Bear and Company.
(Topics):creation spirituality.

Fox, M. (1983b). *Original blessing.* Santa Fe: Bear Publishing Co.
(Topics):creation spirituality.

Fox, T. C. (1987, Sept. 4). Coalition draws belated church into ecosystem. *National Catholic Reporter,* pp. 1+.
Article based on the first North American Conference on Christianity and Ecology meeting on Aug. 19-21, 1987.
(Topics):popular.

Frair, W. (1969). Ignorance, inertia, and irresponsibility. *Journal of the American Scientific Affiliation*, 21(2), 43-44.
(Topics):popular.

Frame, R. (1988). Planetary justice. *Christianity Today*, 32(17), 74.
Report on a forum sponsored jointly by the North American Conference on Christianity and Ecology and the World Alliance of Reformed Churches held in Madison, Wisconsin, in October 1988.
(Topics):popular.

Frame, R. (1990). Christianity and ecology: a better mix than before. *Christianity Today*, 34(April 23), 38-39.
See also interview "Stewardship of the Garden" with Representative Paul Henry (R-Mich.) on the same pages.
(Topics):popular.

Frame, R. (1991). Defiler of the Earth. *Christianity Today*, 35(7):41.
Book review of *Project Earth: Preserving the World God Created* by W. B. Badke.
(Topics):popular.

Francis, J., & Albrecht, P. (Eds.) (1976). *Facing up to nuclear power, risks and potentialities of the large-scale use of nuclear energy*. Philadelphia: Westminister.
(Topics):nuclear.

Francis, M. (1988). Ecology or the integrity of creation. *SEDOS Bulletin*, 11, 378-380.
Discusses the Polonoroeste Project in the state of Rondonia, Brazil.
(Topics):popular, agriculture.

Freeman, J. (1987). Christian conservation begins in the home. In F. W. Krueger (Ed.), *Christian ecology: building an environmental ethic for the twenty-first century* (pp. 63-

64). North Webster, Indiana: The North American Conference on Christianity and Ecology.
(Topics):popular, lifestyle.

French, W. (1986). Technology and ethics: reflections after Chernobyl. *Christian Century*, 103(23), 675-678.
(Topics):technology, ethics, nuclear.

Fretheim, T. E. (1969). *Creation, fall and flood.* Minneapolis: Augsburg Publishing House.
(Topics):theology.

Freudenberger, C. D. (1981). *The gift of land.* Los Angeles: Franciscan Communications.
A large and important work accompanied by film-strips.
(Topics):land, resources, pollution, lifestyle, book.

Freudenberger, C. D. (1982). Resource abuse: "the land does not lie." In W. Byron (Ed.), *The causes of world hunger.* New York: Paulist Press.
(Topics):resources, agriculture.

Freudenberger, C. D. (1984). *Food for Tomorrow?* Minneapolis: Augsburg Publishing House.
(Topics):land, agriculture.

Freudenberger, C. D. (1986). Food and politics: business as usual has run its course. What are the options? *Word & World*, 6(1), 40-53.
(Topics):land.

Freudenberger, C. D. (1987a). Implications of a new land ethic. In B. F. Evans & G. D. Cusack (Eds.), *Theology of the land* (pp. 69-84). Collegeville, MN: The Liturgical Press.
(Topics):land, theology.

Freudenberger, C. D. (1987b). An overview of Christian agriculture historically and principles of land stewardship. In F. W. Krueger (Ed.), *Christian ecology: building an en-*

vironmental ethic for the twenty-first century (pp. 34-35). North Webster, Indiana: The North American Conference on Christianity and Ecology.
(Topics):land, agriculture.

Freudenberger, C. D. (1987c). Regenerative food systems in perspective of the ancient Hebrew tradition of covenant. In F. W. Krueger (Ed.), *Christian ecology: building an environmental ethic for the twenty-first century* (p. 46). North Webster, Indiana: The North American Conference on Christianity and Ecology.
(Topics):land, agriculture.

Freudenberger, C. D. (1988). The agricultural agenda for the twenty-first century. *Pro Rege*, 16(4):15-21.
Pro Rege is published quarterly by the faculty of Dordt College, Sioux Center, Iowa.
(Topic):agriculture.

Freudenberger, C. D. (1989a). The agricultural agenda for the 21st century. *Earthkeeping: A Quarterly on Faith and Agriculture*, 5(1), 5-9.
(Topics):agriculture, appropriate technology, land.

Freudenberger, C. D. (1989b, January - March). Caring for the Earth. *Together, A Publication of World Vision*, pp. 7-9.
(Topics):popular.

Freudenberger, C. D. (1990). *Global dust bowl: can we stop the destruction of the land before it's too late?* Minneapolis: Augsburg.
(Topics):land, agriculture, ethics.

Freudenberger, C. D., & J. C. Hough, Jr. (1977). Lifeboats and hungry people. In D. T. Hessel (Ed.), *Beyond survival: bread and justice in Christian perspective* (pp. 19-47). New York: Friendship Press.
(Topics):ethics.

Freudenstein, E. G. (1970). Ecology and the Jewish tradition. *Judaism*, 19, 406-414.
(Topics):Judaism.

Friend, J. A. (1982). Nature, man and God: a temple revisited. *Reformed Theological Review*, 41(2), 34-41.
(Topics):theology.

Friends (Religious Society of -- Pacific Yearly Meeting): Unity with Nature Committee (1988, June). *Environment in Friends' concerns.*
Write to Environment in Friends' Concerns, 7899 St. Helena Road, Santa Rosa, CA 95404.
(Topics):popular.

Fritsch, A. J. (1972). *A theology of the Earth.* Washington, D. C: CLB Publishers, Inc.
(Topics):theology.

Fritsch, A. J. (1987a). Statement by Fr. Al Fritsch, SJ, at press conference. In F. W. Krueger (Ed.), *Christian ecology: building an environmental ethic for the twenty-first century* (p. 5). North Webster, Indiana: The North American Conference on Christianity and Ecology.
(Topics):popular.

Fritsch, A. J. (1987b). The ethics of direct action. In F. W. Krueger (Ed.), *Christian ecology: building an environmental ethic for the twenty-first century* (p. 69). North Webster, Indiana: The North American Conference on Christianity and Ecology.
(Topics):popular.

Fritsch, A. J. (1987c). *Renew the face of the Earth.* Chicago: Loyola Univ. Press.
(Topics):theology.

Fritsch, A. J., G. McMahan, A. Okagaki, & W. Millard. (1980). *Environmental ethics: choices for concerned citizens.* Garden City, N.Y.: Anchor Press/Doubleday.

(Topics):ethics.

Froehlich, K. (1970). Ecology of creation. *Theology Today*, 27(3), 263-276.
    (Topics):theology.

Froelich, J. (1987a). A Christian ecologist's response to Native American presence and spirituality. In F. W. Krueger (Ed.), *Christian ecology: building an environmental ethic for the twenty-first century* (p. 77). North Webster, Indiana: The North American Conference on Christianity and Ecology.
    (Topics):Indian, Native American.

Froelich, J. (1987b). Technologies for regeneration: part three: the water crisis and you. In F. W. Krueger (Ed.), *Christian ecology: building an environmental ethic for the twenty-first century* (pp. 91-92). North Webster, Indiana: The North American Conference on Christianity and Ecology.
    (Topics):water pollution.

Fuerst, W. J. (1980). Space and place in the Old Testament. *Dialog* (St. Paul, MN, imprint), 19, 193-198.
    (Topics):theology.

Fuggle, R. F. (1987). Convergence between religion and conservation: a review of the Assisi celebrations. In W. S. Vorster (Ed.), *Are We Killing God's Earth?* (pp. 1-5). Pretoria, South Africa: University of South Africa.
    (Topics):theology.

Funck, L. L. (1976). Energy crisis: where are we now? *Moody Monthly*, 76, 118-121.
    (Topics):energy.

Galloway, A. D. (1951). *The cosmic Christ*. New York: Harper and Row.
    (Topics):theology.

Gambell, R. (1990). Whaling -- a Christian position. *Science and Christian Belief,* 2(April), 15-24.
(Topics):popular, natural resources.

Gammie, J. G. (1978). Behemoth and leviathan: on the didactic and theological significance of Job 40:15-41:26. In J. G. Gammie,W. A. Brueggemann,W. L. Humphreys, & J. M. Ward (Eds.), *Israelite wisdom: theological and literary essays in honor of Samuel Terrien* (pp. 217-231). New York: Union Theological Seminary.
(Topics):theology.

Gammie, J. G., Brueggemann, W. A., Humphreys, W. L., & Ward, J. M. (Eds.) (1978). *Israelite wisdom: theological and literary essays in honor of Samuel Terrien.* New York: Union Theological Seminary.
(Topics):theology.

Gardiner, R. W. (1990). Between two worlds: humans in nature and culture. *Environmental Ethics,* 12(4), 339-352.
(Topics):ethics.

Carrick, I. (1970). Right involvement with nature. *Frontier,* 13(2), 31-33.
(Topics):popular.

Garriott, C. (1983, January). A theology for all my relatives: are you at peace with your nonhuman relatives in creation? *Other Side,* pp. 16-19.
(Topics):popular.

Garriott, C. (1987, May). Mound planting: meeting Christ in the garden. *Other Side,* pp. 32-37.
(Topics):agriculture, land.

Gaston, W. F. (1985). *The art of wakefulness: introducing a creation-centered spirituality into the life of the church.* D.Min., United Theological Seminary.
(Topics):theology.

Geels, E. J. (1978). A journal symposium -- the recombinant DNA controversy: examine the dangers and benefits carefully. *Journal of the American Scientific Affiliation*, 30(2), 75-76.
(Topics):genetic engineering.

Geiger, D. R. (1978). Agriculture, stewardship, and a sustainable future. In M. E. Jegen & B. V. Manno (Eds.), *The Earth is the Lord's* (pp. 88-99). New York: Paulist Press.
(Topics):land, agriculture.

Geisler, N. (1989). *Knowing the truth about creation: how it happened and what it means for us.* Ann Arbor, MI: Servant Books.
     Stewardship issues are a minor but important part of the book.
(Topics):popular book.

Gelderloos, O. (1982). Leadership in environmental ethics. In E. R. Squiers (Ed.), *The environmental crisis: the ethical dilemma* (pp. 345-361). Mancelona, MI: Au Sable Institute of Environmental Studies.
(Topics):ethics.

Gelderloos, O. (1989, June). Energy and the Bible. *ICE Melter Newsletter*, p. 1-11.
(Topics):energy, air pollution.

Genesis Rabbah (1985). *Genesis Rabbah: The Judaic Commentary to the Book of Genesis.* Atlanta: Scholars Press.
(Topics):theology, Judaism.

George, D. (1984). *The Christian As a Consumer.* Philadelphia: The Westminster Press.
(Topics):simple lifestyle.

Gibbs, J. C. (1971a). *Creation and redemption: a study in Pauline theology.* Leiden: E. J. Brill.
(Topics):theology.

Gibbs, J. C. (1971b). Pauline cosmic Christology and ecological crisis. *Journal of Biblical Literature*, 90(4), 466-479.
(Topics):theology.

Gibson, W. E. (1975). Economic justice. *Church and society*, 66(November-December), 11-21.
(Topics):ethics, justice.

Gibson, W. E. (1976, Early Winter). Eco-justice: a mission priority for the future. *Connexion* [pages uncertain, not seen].
(Topics):popular.

Gibson, W. E. (1977a). Ecojustice: burning word: Heilbroner and Jeremiah to the church. *Foundations*, 20(4), 318-328.
(Topics):popular, ecojustice.

Gibson, W. E. (1977b). The lifestyle of Christian faithfulness. In D. T. Hessel (Ed.), *Beyond survival*. New York: Friendship Press.
(Topics):lifestyle, popular.

Gibson, W. E. (1978 [revised 1981]). *A covenant group for lifestyle assessment*. New York: Program Agency, Presbyterian Church, U.S.A.
(Topics):lifestyle, popular.

Gibson, W. E. (1979, May). Life on a shrinking planet. *A.D.* [pages unknown, not seen].
(Topics):popular.

Gibson, W. E. (1980a). Justice, the church and the land grant college. In D. T. Hessel (Ed.), *The agricultural mission of churches and land grant universities*. Ames, Iowa: Iowa State University Press.
(Topics):popular.

Gibson, W. E. (1980b). Values for post-affluent times. In D. T. Hessel (Ed.), *Rethinking social ministry.* New York: Program Agency, Presbyterian Church, U.S.A.
  (Topics):popular.

Gibson, W. E. (1980c). What is happening in American society? and the powers of the church. In D. T. Hessel (Ed.), *Participant's manual for social ministry institutes.* New York: Program Agency, Presbyterian Church, U.S.A.
  (Topics):popular.

Gibson, W. E. (1981). Sufficiency: direction of the church's mission. In D. T. Hessel (Ed.), *Congregational lifestyle change for the lean years.* New York: Program Agency, Presbyterian Church, U.S.A.
  (Topics):popular, lifestyle.

Gibson, W. E. (1982a, December). Eco-justice: what is it? *The Egg: A Journal of Eco-Justice* [pages unknown, not seen].
  (Topics):popular, ecojustice.

Gibson, W. E. (1982b, September). The ecology of an accelerating transition. *The Egg: A Journal of Eco-Justice* [pages unknown, not seen].
  (Topics):popular.

Gibson, W. E. (1982c, July-August). Shalom through work. *The Upper Room* [pages unknown, not seen].
  (Topics):popular.

Gibson, W. E. (1982d). *Social ministry and the knowledge of God in our time.* Ithaca, New York: CRESP.
  (Topics):popular.

Gibson, W. E. (1983a, March). The land mourns. *JSAC Grapevine* [pages unknown, not seen].
  (Topics):popular, land.

Gibson, W. E. (1983b). Praise the creator -- care for creation. In D. T. Hessel (Ed.), *Social themes of the Christian year: a commentary on the lectionary.* Philadelphia: The Geneva Press.
(Topics):popular.

Gibson, W. E. (1985a). Eco-justice: new perspective for a time of turning. In D. Hessel (Ed.), *For Creation's sake: preaching, ecology and justice* (pp. 15-31). Philadelphia: Geneva Press.
(Topics):theology, popular, ecojustice.

Gibson, W. E. (1985b). The search for a new humility. In N. Hundertmark (Ed.), *Pro-earth.* New York: Friendship Press.
(Topics):popular.

Gibson, W. E. (1985c). Stewardship and economics. In N. Hundertmark (Ed.), *Pro-earth.* New York: Friendship Press.
(Topics):economics, popular, stewardship.

Gibson, W. E. (1986a). Confessing and Covenanting. In D. T. Hessel (Ed.), *Shalom connections in personal and congregational life.* Ellenwood Georgia: Alternatives.
(Topics):popular.

Gibson, W. E. (1986b, Spring). Three norms for our time. *The Egg: A Journal of Eco-Justice* [pages unknown, not seen].
(Topics):popular.

Gibson, W. E. (1987a, Spring). Creation and liberation as a continuing story. *The Egg: A Journal of Eco-Justice* [pages unknown, not seen].
(Topics):Popular.

Gibson, W. E. (1987b). Sustainable security: peace education and ecological/economic responsibility. In D. T. Hessel

(Ed.), *Moving toward shalom: essays in memory of John T. Conner*. Nyack, New York: Fellowship of Reconciliation. (Topics):economics, popular.

Gibson, W. E. (1988). An order in crisis, and the declaration of new things. In R. L. Stivers (Ed.), *Economics and the reformed faith*. Lanham, MD: University Press of America. (Topics):economics, popular.

Gibson, W. E. (1988/89). Valuing Earth: probing the why and the how. *The Egg: A Journal of Eco-Justice*, 8(4), 1-2. (Topics):land.

Gibson, W. E. (1989a). Hearings reveal toxic contamination of minority communities. *The Egg: A Journal of Eco-Justice*, 9(3), 13-15. (Topics):pollution and toxic chemicals.

Gibson, W. E. (1989b). Managing waste and doing justice. *The Egg: A Journal of Eco-Justice*, 9(2), 1. (Topics):solid waste, hazardous waste, pollution.

Gibson, W. E. (1989c). The universe as our home -- but not ours alone. *The Egg: A Journal of Eco-Justice*, 9(3), 2. (Topics):popular.

Gibson, W. E. (1990). Beginning a "turnaround decade"? the many faces of Earth Day. *Christianity and Crisis: a Christian Journal of Opinion*, 50(May 14), 147-149. (Topics):popular.

Giles, F. H., Jr. (1974). No uniquely Christian response. *Journal of the American Scientific Affiliation*, 26(1), 16-17. (Topics):population.

Gilkey, L. (1954). *Maker of heaven and earth: a thesis on the relation between metaphysics and Christian theology with special reference to the problem of creation as that problem appears in the philosophies of F. H. Bradley and A. N.*

Whitehead and in the historic leaders of Christian thought.
Ph.D., Columbia University.
    (Topics):theology.

Gilkey, L. (1965). *Maker of heaven and earth.* Garden City,
N.Y.: Doubleday.
    (Topics):theology.

Gilkey, L. (1974). The theological understanding of humanity and nature in a technological era. *Anticipation,*
19(November), 33-35.
    (Topics):popular, technology.

Gilkey, L. (1987). Is religious faith possible in an age of
science? In F. T. Birtel (Ed.), *Religion, science, and public
policy* (pp. 49-64). New York: The Crossroads Publishing
Company.
    (Topics):popular.

Gill, J. H. (1978, May). The ethics of environment. *The
Reformed Journal,* pp. 18-21.
    (Topics):ethics.

Gillette, R. (1973). Puerto Rico: pollution and development.
Church and Society, 64(2), 16-24.
    (Topics):pollution, popular, energy.

Gingerich, O. (1974). Selfish motivation. *Journal of the
American Scientific Affiliation,* 26(1), 17.
    (Topics):population.

Gish, A. (1977, November). Our little shovels and rakes.
Other Side, pp. 26-27.
    (Topics):popular.

Gittens, A. J. (1986). Ecology and world poverty: a Christian
response. *Spirituality Today,* 38, 19-30.
    (Topics):lifestyle.

Glacken, C. J. (1967). *Traces on the Rhodian Shore: nature and culture in western thought from ancient times to the end of the eighteenth century.* Berkeley: Univ. of Calif. Press.
    A monumental work.
    (Topics):history.

Glacken, C. J. (1970a). Man against nature: an outmoded concept. In H. W. Helfrich, Jr. (Ed.), *The environmental crisis: man's struggle to live with himself.* New Haven, Conn.:Yale University Press.
    (Topics):history.

Glacken, C. J. (1970b). Man's place in nature in recent western thought. In M. Hamilton (Ed.), *This little planet* (pp. 163-202). New York: Charles Scribner's Sons.
    (Topics):history.

Gladwin, J. (1979). *God's people in God's world.* Downers Grove, IL: IVP.
    See especially ch. 3.
    (Topics):book.

Glover, S. (1990). Farming in faith: when an accident forced the Bouwman family to restructure their farm, they decided to change their method of farming as well. *Earthkeeping: a quarterly on faith and agriculture,* 6(1):8-9.
    (Topics):land, agriculture.

Gomes, P. J. (1977). Critique of W. E. Gibson's "Ecojustice: a burning word." *Foundations,* 20(4), 329-332.
    (Topics):popular, ecojustice.

Gonzalas, P. (1987). Technologies for regeneration: part one: the house that junk built. In F. W. Krueger (Ed.), *Christian ecology: building an environmental ethic for the twenty-first century* (p. 90). North Webster, Indiana: The North American Conference on Christianity and Ecology.
    (Topics):energy (solar), appropriate technology.

Goodwin, R. D., Jr. (1976). *Toward an ecological awareness: a theology of ecology and the role of the church.* D.Min., Drew University.
(Topics):theology.

Goossen, G. (1987). A "boomerang" environmental theology. *St. Mark's Review,* 131, 32-42.
(Topics):theology, resources, St. Francis of Assisi, land.

Gordis, R. (1970). The Earth is the Lord's -- Judaism and the spoilation of nature. *Keeping Posted,* 16(3), 5-9.
(Topics):Judaism.

Gordis, R. (1971, April 2). Judaism and the spoilation of nature. *Congress Bi-Weekly.*
(Topics):Judaism.

Gordis, R. (1985). Job and ecology (and the significance of Job 40:15). *Hebrew Annual Review,* 9, 189-202.
(Topics):Judaism, theology.

Gosling, D. (1986). Towards a credible ecumenical theology of nature. *Ecumenical Review,* 38(3), 322-331.
(Topics):theology.

Gottwald, N. K. (1985). The Biblical mandate for eco-justice action. In D. Hessel (Ed.), *For Creation's sake: preaching, ecology and justice* (pp. 32-44). Philadelphia: Geneva Press.
(Topics):theology, ecojustice.

Goudzwaard, B. (1987). Creation management: the economics of Earth stewardship (Part one). *Epiphany,* 8(Fall), 37-45.
(Topics):economics.

Goudzwaard, B. (1988). Creation management: the economics of Earth stewardship (Part two). *Epiphany,* 8(2), 67-72.
(Topics):economics.

Govier, G. (1989). Take the Bible outdoors. *Moody Monthly*, 90(2), 17.
> Focus is on Dr. Susan Bratton and her work in caring for creation. Dr. Bratton is a plant ecologist and research coordinator for the National Park Service. She is on the faculty at the University of Georgia Institute of Ecology and Au Sable Institute of Environmental Studies.
> (Topics):popular.

Gowan, D. E. (1970). Genesis and ecology: does "subdue" mean "plunder"? *Christian Century*, 87(40), 1188-1191.
> (Topics):popular.

Gowan, D. E. (1973). Ecotheology; a review article. *Perspective* , 14, 107-113.
> Pittsburgh Theological Seminary.
> (Topics):theology.

Gowan, D. E. (1985). The fall and redemption of the material world in apocalyptic literature. *Horizons in Biblical Theology: An International Dialogue*, 7(2), 83-103.
> (Topics):theology.

Gowan, D. E., & Schumaker, M. (1973). *Genesis and ecology: an exchange*. Kingston, Ontario: Queen's Theological College.
> Gowan 1970 is reprinted. This is responded to by Schumaker in an article titled *Nature's servant-King* (pp. 14-26). A rejoinder by Gowan appears on pp. 27-31. The work was published as the first paper in the Occasional Papers series.
> (Topics):popular.

Graef, S. (1990). The foundation of ecology in Zen Buddhism. *Religious Education*, 85(1), 42-50
> (Topics):Buddhism.

Graham, R. A. (1990). The environment and the gospel. *Columbia*, 70(6), 22.

(Topics):popular.

Granberg-Michaelson, W. (1981a). At the dawn of the new creation: a theology of the environment. *Sojourners*, 10(11), 13-16.
(Topics):popular.

Granberg-Michaelson, W. (1981b, October 16). More biblical than Watt. *The Church Herald.*
(Topics):popular.

Granberg-Michaelson, W. (1981c, October 30). Toward a theology of the Earth. *The Church Herald*, p. 4-7.
(Topics):popular.

Granberg-Michaelson, W. (1982a, November 12). Called to be caretakers. *The Church Herald*, p. 8-10.
(Topics):popular.

Granberg-Michaelson, W. (1982b). The Earth is the Lord's. *Radix*, 14(3), 3-4.
(Topics):popular.

Granberg-Michaelson, W. (1982c). Earthkeeping: a theology for global sanctification. *Sojourners*, 11(9), 21-24.
(Topics):popular.

Granberg-Michaelson, W. (1982d). The ethics of strip-mining coal in Montana. In E. R. Squiers (Ed.), *The environmental crisis: the ethical dilemma* (pp. 293-308). Mancelona, MI: Au Sable Institute of Environmental Studies.
(Topics):ethics, energy.

Granberg-Michaelson, W. (1983a). The authorship of life: a biblical look at genetic engineering. *Sojourners*, 12(6), 18-22.
(Topics):genetic engineering.

Granberg-Michaelson, W. (1983b). Farming with justice. *The Other Side*, 19(3), 38-43.
 (Topics):land.

Granberg-Michaelson, W. (1984a). The promise of God's reign: the church's role in the world's future. *Sojourners*, 13(6), 16-19.
 (Topics):popular.

Granberg-Michaelson, W. (1984b). Redeeming the Earth: a theology for this world. *Covenant Quarterly*, 42(2), 17-29.
 (Topics):theology.

Granberg-Michaelson, W. (1984c). *A worldly spirituality: the call to redeem life on Earth*. San Francisco: Harper & Row, Publishers.
 (Topics):book.

Granberg-Michaelson, W. (1985, January). Caring for God's Earth. *Response*.
 (Topics):popular.

Granberg-Michaelson, W. (1986, January). Biblical perspectives on life and death. *Perspectives*.
 (Topics):popular.

Granberg-Michaelson, W. (1987a, September). Fly-fishing as a spiritual discipline. *Perspectives*.
 (Topics):popular.

Granberg-Michaelson, W. (1987b). Introduction: identification or mastery? In W. Granberg-Michaelson (Ed.), *Tending the garden: essays on the gospel and the earth* (pp. 1-5). Grand Rapids: Eerdmans Publishing Company.
 (Topics):popular.

Granberg-Michaelson, W. (Ed.) (1987c). *Tending the garden: essays on the gospel and the Earth*. Grand Rapids: Eerdmans.
 (Topics):book, theology.

Granberg-Michaelson, W. (1987d, June). This creation and the new creation. *Perspectives*.
(Topics):popular.

Granberg-Michaelson, W. (1988a). *Ecology and life: accepting our environmental responsibility*. Waco, TX: Word Books.
The second volume in the Issues of Christian Conscience series edited by Vernon Grounds.
(Topics):book.

Granberg-Michaelson, W. (1988b, May 6). Rediscovering our call to God's creation. *The Church Herald*, pp. 12-14.
(Topics):popular.

Granberg-Michaelson, W. (1988c). Why Christians lost an environmental ethic. *Epiphany*, 8(2), 40-50.
Paper presented at NACCE Conference in 1988.
(Topics):popular, history.

Granberg-Michaelson, W. (1989a). Interview by Judith Hougen. *The Other Side*, 25, 15-19.
(Topics):popular.

Granberg-Michaelson, W. (1989b). Why Christians lost an environmental ethic. *Firmament: The Quarterly of Christian Ecology*, 1(3), 4-7.
(Topics):history.

Granberg-Michaelson, W. (1990). Renewing the whole creation: constructing a theology of relationship. *Sojourners*, 19(F/Mr), 10-14.
(Topics):theology, popular.

Grandy, J. W., & Ragan, P. H. (1989). Humans and animals: drawing lines with conscience. *The Egg: A Journal of Eco-Justice*, 9(3), 7-9.
(Topics):popular.

Grant, C. (1986). Humanity, nature, God. *Grail: An Ecumenical Journal*, 2, 7-15.
　　(Topics):popular.

Gray, E. D. (1981). *Green paradise lost*. Wellesley, MA: Roundtable Press.
　　Combines feminist and ecological concerns within the Judeo-Christian tradition.
　　(Topics):book, ecofeminist.

Gray, E. D. (1985). A critique of dominion theology. In D. Hessel (Ed.), *For Creation's sake: preaching, ecology and justice* (pp. 71-83). Philadelphia: Geneva Press.
　　(Topics):theology.

Green, J. (1985). *God's fool: the life and times of Francis of Assisi*. San Francisco: Harper and Row, Publishers.
　　English translation of 1983 French publication.
　　(Topics):book, St. Francis of Assisi.

Green, M. (1988). A living creation -- how shall we love it? *The Egg: A Journal of Eco-Justice*, 8(3), 4-5.
　　(Topics):popular.

Green, S. (1990). Creation Groans? *Impact*, 47(30), 6-9.
　　Published by the Conservative Baptist Foreign Mission Society.
　　(Topics):popular.

Greenoak, F. (1985). *God's acre: the flowers and animals of the parish churchyard*. New York: E. P. Dutton.
　　(Topics):book, popular.

Gregorios, P. (1978). *The human presence: an orthodox view of nature*. *World Council of Churches, Geneva*. Post Box 501, Park Town, Madras, India　600 003: The Christian Literature Society. Published in 1980.
　　(Topics):theology.

Gregorios, P. (1987a). *The human presence: ecological spirituality and the age of the spirit.* Amity, NY: Amity House.
(Topics):theology.

Gregorios, P. (1987b). New Testament foundations for understanding the Creation. In W. Granberg-Michaelson (Ed.), *Tending the garden: Essays on the Gospel and the Earth* (pp. 83-92). Grand Rapids, MI: Eerdmans Publishing Company.
(Topics):theology, popular.

Griffin, D. R. (1973a). A new vision of nature. In *Earth ethics for today and tomorrow: responsible environmental trade-offs.* Bowling Green, OH: Bowling Green State University. (Topics):ethics.

Griffin, D. R. (1973b). Whitehead's contributions to a theology of nature. *The Bucknell Review.*
Was unable to locate. Pages unknown.
(Topics):theology.

Griffioen-Drenth, M. (1987). The soil is the Lord's. *Earthkeeping: A Quarterly on Faith and Agriculture,* 3(5), 3.
(Topics):land.

Griffiths, R. (1982). *The human use of animals.* Bramcote, Notts.: Grove.
(Topics):animal rights.

Gruner, R. (1975). Science, nature, and Christianity. *Journal of Theological Studies,* 26(1), 55-81.
(Topics):history.

Gurnett, J. (1990). The vineworker's farm: waterless toilets and alternative housing are just two of the innovations that make this farm for Bible students unique. *Earthkeeping,* 6(1), 20-21.
(Topics):agriculture.

Gustafson, J. M. (1977). Interdependence, finitude, and sin: reflections on scarcity. *The Journal of Religion*, 57, 156-168.
(Topics):ethics.

Gustafson, J. M. (1981). *Ethics from a theocentric perspective. Vol. I, theology and ethics.* Chicago: University of Chicago Press.
See especially chapter five, pp. 271 ff.
(Topics):ethics, theology.

Gustafson, J. M. (1984). *Ethics from a theocentric perspective. Vol. II, ethics and theology.* Chicago: University of Chicago Press.
(Topics):ethics, theology.

Gustafson, J. M. (1985). A response to critics. *Journal of Religious Ethics*, 13, 185-209.
(Topics):theology, ethics.

Gustafson, J. W. (1991). Integrated holistic development and the world mission of the church. In G. T. Prance & C. B. DeWitt (Eds.), *Missionary earthkeeping*. Macon, Georgia: Mercer University Press.
(Topics):popular.

Guttmacher, A. F. (1972). Population and pollution. *Review and Expositor*, 69(1), 55-65.
(Topics):population, pollution.

Haas, J. W., Jr. (1978). A journal symposium -- the recombinant DNA controversy: dangers less serious than earlier believed. *Journal of the American Scientific Affiliation*, 30(2), 74.
(Topics):genetic engineering.

Habel, N. (1972). Yahweh, maker of heaven and earth. *Journal of Biblical Literature*, 91, 321-337.
(Topics):theology.

Habig, M. (Ed.) (1973). *The omnibus of sources*. Chicago: The Franciscan Herald Press.
(Topics):Saint Francis of Assisi, theology.

Haenke, D. (1987). Bioregionalism: the natural lines of creation. In F. W. Krueger (Ed.), *Christian ecology: building an environmental ethic for the twenty-first century* (p. 54). North Webster, Indiana: The North American Conference on Christianity and Ecology.
(Topics):popular.

Haenke, D. (1988). NACCE beginnings. *Firmament: The Quarterly of Christian Ecology*, 1(1), 5.
(Topics):popular.

Hahm, P. (1972, September). A Korean comment on the Christian concern for the future of the global environment. *Anticipation*, pp. 24ff.
(Topics):popular.

Hall, D. J. (1982). *The steward: a biblical symbol come of age*. New York: Friendship Press.
(Topics):stewardship, theology.

Hall, D. J. (1985). *Christian mission: the stewardship of life in the kingdom of death*. New York: Friendship Press.
(Topics):stewardship, theology.

Hall, D. J. (1986). *Imaging God: dominion as stewardship*. New York: Eerdmans/Friendship Press -- for the Commission of Stewardship of the National Council of the Churches of Christ in the U.S.A.
(Topics):theology, stewardship.

Hallman, D. 1989. *Caring for creation*. Winfield, Canada: Wood Lake Books.
(Topics):popular book.

Halteman, J. (1982). Conservation and its short run effects on distribution in bottleneck resource industries. In E. R.

Squiers (Ed.), *The environmental crisis: the ethical dilemma* (pp. 279-292). Mancelona, MI: Au Sable Institute of Environmental Studies.
(Topics):resources.

Halver, J. E. (1989). Are we responsible for the Earth? *Decision*, 30(11), 14-15.
Lay audience. Published by Billy Graham Evangelistic Association.
(Topics):stewardship, Billy Graham, popular.

Hamilton, M. (Ed.) (1970). *This little planet*. New York: Charles Scribner's Sons.
(Topics):popular book.

Hammer, G. D. (1976). Response to James Donaldson. *Foundations: A Baptist Journal of History and Theology*, 19(3), 254ff.
(Topics):popular.

Hammerton, H. J. (1967). Unity of creation in the apocalypse. *Church Quarterly Review*, 168, 20-33.
(Topics):theology.

Hand, J. A. (1969) *Teleological aspects of creation: a comparision of the concepts of being and meaning in the theologies of Jonathan Edwards and Paul Tillich*. Ph.D., Vanderbilt University.
(Topics):theology.

Hanson, J. (1986). The long-term energy crisis: "a little breathing room" but wise choices demanded. *Engage/Social Action*, 14(4), 4-10.
(Topics):energy.

Harakas, S. S. (1988). The integrity of creation and ethics. *Saint Vladimir's Theological Quarterly*, 32(1), 27-42.
(Topics):ethics.

Harblin, T. D. (1977). Mine or garden? Values and the environment -- probable sources of change in the next hundred years. *Zygon*, 12(2), 134-150.
(Topics):ethics.

Harder, A. (1971). Ecology, magic and the death of man. *Christian Scholar's Review*, 1(2), 117-131.
(Topics):popular.

Hardin, G. (1968). The tragedy of the commons. *Science*, 162, 1243-1248.
Important scientific paper, not with Christian emphasis. Portions reprinted in *Journal of the American Scientific Affiliation* (1969), Vol. 21(3):83-87.
(Topics):resources.

Hardin, G. (1980). Ecology and the death of providence. *Zygon*, 15(1), 57-69.
(Topics):popular.

Hareuveni, N. (1980). *Nature in our biblical heritage*. Kiryat Ono, Israel: Neot Kedumim, Ltd.
(Topics):Holy Land, Judaism, Israel, scriptural plants and animals.

Hareuveni, N. (1984). *Tree and shrub in our biblical heritage*. Kiryat Ono, Israel: Neot Kedumim Ltd.
(Topics):Holy Land, Israel, Judaism, scriptural plants and animals.

Hargrove, E. C. (1979). The historical foundations of American environmental attitudes. *History*, 1(3), 209-240.
Excellent review, but not emphasizing the Christian perspective.
(Topics):ethics.

Hargrove, E. C. (Ed.) (1986). *Religion and environmental crisis*. Athens, GA: University of Georgia Press.
An important collection of papers centering on the topic of environmental ethics. In addition to the Christian

World View, other papers examine the influence of Polytheism, Native American religions, Judaism, Taoism, and Islam in shaping environmental attitudes.
(Topics):Judaism, Islam, ethics.

Harrison, B. W., Stivers, R. L., & Stone, R. H. (Eds.) (1986). *The public vocation of Christian ethics*. New York: The Pilgrim Press.
(Topics):ethics.

Hart, A. (1989). "Where the river flows": ecology and the orthodox liturgy. *Epiphany*, 9, 31-36.
(Topics):popular.

Hart, J. (1984). *The spirit of the earth: A theology of the land*. New York: Paulist Press.
(Topics):book, land, theology.

Hart, J. (1987). Land, theology, and the future. In B. F. Evans & G. D. Cusack (Eds.), *Theology of the land* (pp. 85-102). Collegeville, MN: The Liturgical Press.
(Topics):land, theology.

Hatfield, M. (1977). *Between a rock and a hard place*. Waco,TX: Word Incorporated, Pocket Book Edition.
See especially Ch. 12.
(Topics):popular.

Haught, J. F. (1989). Religion and cosmic homelessness: some environmental implications. *The Egg: A Journal of Eco-Justice*, 9(3), 4-7.
(Topics):popular.

Hawksley, R. L. (1987). Redefining the family in "family farm": ecological justice and the biblical call to community. In F. W. Krueger (Ed.), *Christian ecology: building an environmental ethic for the twenty-first century* (pp. 44-45). North Webster, Indiana: The North American Conference on Christianity and Ecology.
(Topics):land, agriculture.

Hay, D. A. (1990). Christians in the global hothouse: Tyndale Fellowship annual lecture in ethics. *Tyndale Bulletin*, 41(May), 109-127.
(Topics):ethics, pollution (greenhouse effect).

Hayeland, J. (1971). Of beer cans and bleach bottles. *Religion Teacher's Journal*, 5, 17-20.
(Topics):popular.

Hayes, Z. (1980). *What are they saying about creation?* New York: Paulist Press.
(Topics):theology.

Head, L., and Guerrero, M. (1990). Environmental racism and the struggle for justice. *The Witness*, 73(September), 8-10.
(Topics):pollution.

Hearn, W. (1974). Stewardship begins with us. *Journal of the American Scientific Affiliation*, 26(1), 17-18.
(Topics):population.

Hearn, W. (1982). Signs of hope. *Radix*, 14(3), 4-9.
(Topics):popular.

Hebblethwaite, P. (1988, Nov. 4). Vatican statement on ecology possible, overdue. *National Catholic Reporter*, p. 4.
(Topics):popular.

Heddendorf, R. (1974). Individual rights vs. common good. *Journal of the American Scientific Affiliation*, 26(1), 18.
(Topics):population.

Hefley, J. (1970). Christians and the pollution crisis. *Moody Monthly*, 71(9), 18-21.
(Topics):pollution.

Hefner, P. J. (Ed.) (1964). *The scope of grace: essays on nature and grace in honor of Joseph Sittler*. Philadelphia: Fortress Press.
(Topics):popular, theology.

Hefner, P. J. (1969). Towards a new doctrine of man: the relationship of man and nature. In B. E. Meland (Ed.), *The future of empirical theology* (pp. 235-266). Chicago: University of Chicago Press.
(Topics):theology.

Hefner, P. J. (1974). The politics and ontology of nature and grace. *Journal of Religion*, 54(April), 138-153.
(Topics):theology.

Heideman, E. (1986). Beyond dung: a theology of manure. *Third Way*, 9(2), 24-26.
(Topics):land, theology.

Heinegg, P. (1976). Ecology and the fall. *The Christian Century*, 93(17), 464-466.
(Topics):popular.

Heiss, R. L., & McInnis, N. F. (Eds.) (1971). *Can man care for the Earth?* New York: Abingdon Press.
(Topics):book.

Helfand, J. I. (1971). Ecology and the Jewish tradition: a postscript. *Judaism: A Quarterly Journal of Jewish Life and Thought*, 20, 330-335.
(Topics):Judaism.

Helgeland, J. (1980). Land and eschatology. *Dialog*, 19, 186-192.
(Topics):land, theology.

Helms, J. D. (1984). Walter Lowdermilk's journey: from forester to land conservationist. *Environmental Review*, 8(Summer):133-145.
(Topics):popular.

Helmuth, K. (1987). Sustainable agriculture and the church: a panel presentation: part one. In F. W. Krueger (Ed.), *Christian ecology: building an environmental ethic for the twenty-first century* (pp. 47-48). North Webster, Indiana: The North American Conference on Christianity and Ecology.
(Topics):land, agriculture.

Hendricks, W. L. (1971). Stewardship in the New Testament. *Southwestern Journal of Theology*, 13(2), 25-33.
(Topics):stewardship, theology.

Hendry, G. S. (1971). The eclipse of creation. *Theology Today*, 28(4), 406-425.
(Topics):theology.

Hendry, G. S. (1973). Consider the lilies. *Princeton Seminary Bulletin*, 66(October), 25-32.
(Topics):theology.

Hendry, G. S. (1980). *Theology of nature.* Philadelphia: Westminster Press.
(Topics):theology.

Henrey, K. H. (1954-55). Land Tenure in the Old Testament. *Palestine Exploration Quarterly*, 55, 5-15.
(Topics):theology, land.

Henriot, P. (1980). Why the eighties scare me. *Other Side*, 16, 26-32.
(Topics):popular.

Henry, C. F. H. (Ed.) (1978). *Horizons of Science.* New York: Harper and Row.
See pages 63-86 in the chapter, "Environmental problems and the Christian ethic."
(Topics):popular, ethics.

Henry, C. F. H. (1987a). Keeping the Earth: a theology. *Eternity, 38*(7/8), 13.
(Topics):popular.

Henry, C. F. H. (1987b). Stewardship of the environment. In K. S. Kantzer (Ed.), *Applying the scriptures: papers from ICBI summit III* (pp. 473-488). Grand Rapids, MI: Academie Books, Zondervan Publishing House.
(Topics):theology, stewardship.

Hens-Piazza, G. (1983). A theology of ecology: God's image and the natural world. *Biblical Theology Bulletin, 13*, 107-110.
(Topics):theology.

Hermisson, H. (1978). Observations on the creation theology in wisdom. In J. G. Gammie, W. A. Brueggemann, W. L. Humphreys, & J. M. Ward (Eds.), *Israelite wisdom: theological and literary essays in honor of Samuel Terrien* (pp. 43-57). New York: Union Theological Seminary.
(Topics):theology.

Herrmann, R. L. (1976). Human engineering and the Church. *Journal of the American Scientific Affiliation, 28*(2), 59-62.
(Topics):genetic engineering.

Herrmann, R. L. (1978). A journal symposium -- the recombinant DNA controversy: could anything but good come out? *Journal of the American Scientific Affiliation, 30*(2), 73-74.
(Topics):genetic engineering.

Herrmann, R. L., & Templeton, J. M. (1985). The vast unseen and the genetic revolution. *Journal of the American Scientific Affiliation, 37*(3), 132-141.
(Topics):genetic engineering.

Herron, R. B. (1986). The land, the law, and the poor. *Word & World, 6*(1), 76-84.
(Topics):land, theology.

Herscovici, A. (1985). *Second nature: the animal rights controversy.* Toronto: CBC Enterprises.
(Topics):animal rights.

Hertz, K. H. (1970). Ecological planning for metropolitan regions. *Zygon,* 5(4), 290-303.
Focus is on regional planning in urban areas.
(Topics):land.

Hessel, D. T. (Ed.) (1977). *Beyond survival: bread and justice in Christian perspective.* New York: Friendship Press.
(Topics):land.

Hessel, D. T. (Ed.) (1979). *Energy ethics: a Christian response.* New York: Friendship Press.
(Topics):energy, ethics.

Hessel, D. T. (1985). *For creation's sake: preaching, ecology, and justice.* Philadelphia: Geneva Press.
(Topics):book, ecojustice.

Hessel, D. T. (1986a). For creations's sake: preaching, ecology and justice. *Faith Missionary,* 3(2), 92-94.
(Topics):popular, ecojustice.

Hessel, D. T. (Ed.) (1986b). *Shalom connections in personal and congregational life.* Ellenwood, GA: Alternatives -- copyright, The Program Agency, Presbyterian Church (U.S.A.).
(Topics):popular book.

Hessel, D. T., & Wilson, G. (1981). *Congregational lifestyle changes for the lean years.* New York: United Presbyterian Program Agency.
(Topics):lifestyle.

Hexham, I., & Poewe-Hexham, K. (1988). The soul of the New Age. *Christianity Today,* 32(12), 17-21.

(Topics):New Age.

Hicks, R. L. (1954) *Land (eretz) and (adamah) earth with the Yahwist: A philological, literary, and theological investigation of the concept of the land in Genesis 2:4B-24:7.* Th.D., Union Theological Seminary in the City of New York.
Could not locate a copy to confirm the citation.
(Topics):land, theology.

Hiebert, T. (1990). Ecology and the Bible. *Daughters of Sarah,* 16(3), 12-13.
(Topics):popular.

Hiers, R. H. (1984). Ecology, biblical theology, and methodology: biblical perspectives on the environment. *Zygon,* 19, 43-59.
(Topics):theology.

Higgins, J. (1980). Our plundered planet: is the Bible to blame? *Liguorian,* 68, 20-23.
(Topics):popular.

Higgins, R. (1978). *The seventh enemy: the human factor in the global crisis.* New York: McGraw-Hill Book Company.
See especially chapter 18, titled "The seventh lamp: a re-awakening to the religious."
(Topics):book.

Hodel, D. P. (1987). Leading steward of our resources. *Eternity,* 38(7/8), 15.
Interview by Anita M. Smith.
(Topics):resources.

Hodgson, P. E. (1987). The Christian dimensions of energy problems. *Month,* 20, 73-76.
(Topics):energy, nuclear.

Hoekema, D. A. (1979, August). Does the nuclear option make sense? *The Reformed Journal,* pp. 18-19.

(Topics):nuclear.

Hoenig, S. B. (1968-69). Sabbatical years and the year of jubilee. *The Jewish Quarterly Review*, 59, 222-236.
(Topics):Judaism.

Hogan, R. (1987). Technologies for regeneration: part two: the Christian eco-village. In F. W. Krueger (Ed.), *Christian ecology: building an environmental ethic for the twenty-first century* (p. 91). North Webster, Indiana: The North American Conference on Christianity and Ecology.
(Topics):popular, appropriate technology.

Holmgren, V. (1972). *Birdwalk through the Bible*. New York: The Seabury Press.
Reprinted by Dover Publications, Inc., in 1988.
(Topics):scriptural plants and animals.

Hooykaas, R. (1961). New responsibility in a scientific age. *Free University Quarterly*, 8, 78-97.
(Topics):history.

Horne, B. (1983). *A world to gain: incarnation and the hope of renewal*. London: Darton, Longman and Todd.
See especially Ch. 3.
(Topics):book.

Hough, J. C., Jr. (1980). Land and people: the eco-justice connection. *Christian Century*, 97(30), 910-914.
(Topics):land, ecojustice.

Hough, J. C., Jr. (1981). The care of the Earth: the moral basis for land conservation. *Quarterly Review: A Scholarly Journal for Reflection on Ministry*, 1, 3-22.
(Topics):land.

Houston, J. M. (1972). The environmental movement; five causes of confusion. *Christianity Today*, 16(24), 1130-1132.
(Topics):popular.

Houston, J. M. (1980). *I believe in the creator.* Grand Rapids, MI: Eerdmans.
(Topics):book.

Houston, T. (1986, July-September). Water and community. *Together: A Journal of World Vision International,* p. 1.
(Topics):resources (water).

Hovey, G. (1990). Questions after Earth Day [editorial]. *Christianity and Crisis,* 50(7), 139-140.
(Topics):popular.

Hoyer, G. W. (1971). The environmental crisis and Christian responsibility: introduction. *Concordia Journal,* 42(March), 176-177.
(Topics):popular.

Htun, N. (1987). The state of the environment today: the needs for tomorrow. In S. Davies (Ed.), *Tree of life: Buddhism and protection of nature with a declaration on environmental ethics from His Holiness the Dalai Lama and an introduction by Sir Peter Scott* (pp. 19-30). Buddhist Perception of Nature.
(Topics):Buddhism.

Huber, W. (1988). Justice, peace and the integrity of creation: a challenge for ecumenical theology. *Scriptura* (24), 1-16.
(Topics):theology.

Hughart, T. A. (1982) *A design for teaching a Christian ethic with an ecological world-view.* S.T.D., San Francisco Theological Seminary.
(Topics):theology.

Hughes, J. D. (1975). *Ecology in ancient civilizations.* Albuquerque: University of New Mexico Press.
(Topics):history.

Hulteen, B. (1990a). The cry of Creation [editorial]. *Sojourners*, 19(2):4.
(Topics):popular.

Hulteen, B. (1990b). The greenwashing of America. *Sojourners*, 19(7), 5-6.
(Topics):popular.

Hulteen, B., and Jaudon, B. (1990). With heart and hands: living in God's covenant on a farm. Interview with Richard Cartwright Austin. *Sojourners*, 19(F/Mr):26-29.
(Topics):agriculture, land.

Hultgren, A. J. (Ed.) (1984). Cosmos and creation, thematic issue. *Word and World: Theology for Christian Ministry*, 4(fall): pages ?
(Topics):popular.

Hultgren, A. J. (1986a). Land values and ministry. *Word and World: Theology for Christian Ministry*, 6(1), 3-4.
(Topics):land.

Hultgren, A. J. (Ed.) (1986b). The land, thematic issue. *Word and World: Theology for Christian Ministry*, 6(1), 3-96.
(Topics):land.

Humphrey, C. (Ed.) (1990). [Religious attitudes toward nature.] *ARC: The Journal of the Faculty of Religious Studies, McGill University*, 18(Spring), 5-128.
(Topics):popular.

Hundertmark, N. (1985). *Pro-Earth*. New York: Friendship Press.
(Topics):popular book.

Hunt, D. (1983). *Peace, prosperity and the coming holocaust*. Eugene, OR: Harvest House.

Critical attack on the New Age movement and includes those concerned with care for the creation as part of the New Age.
(Topics):popular book, New Age.

Hunt, S. (1989). The alternative economics movement. *The Egg: A Journal of Eco-Justice*, 9(1), 7-9.
(Topics):economics.

Hurst, J. S. (1972). Towards a theology of conservation. *Theology*, 75, 197-205.
(Topics):theology.

Hutten, A. (1987). Carving out a living organically. *Earthkeeping: A Quarterly on Faith and Agriculture*, 3(5), 19-20.
(Topics):land.

Ice, J. L. (1975). Ecological crisis: radical monotheism vs. ethical pantheism. *Religion in Life*, 44, 203-211.
Discusses Albert Schweitzer's views and suggests that the ethical pantheism of Schweitzer may lead us "beyond the present ecological crisis" (p. 210).
(Topics):popular.

Imsland, D. (1971). *Celebrate the Earth*. Minneapolis: Augsburg Pub. House.
(Topics):book.

Inglis, T. (1987). An ecological theology. *Doctrine and Life*, 37, 74-79.
(Topics):popular.

Ingram, K. J. (1990). Break forth together into singing: a Christian feminist broods over her planet. *Daughters of Sarah*, 16(3),28-30.
(Topics):ecofeminism.

Innes, K. (1987a). Bibliography: Christian attitudes to the environment and human responsibility for it. *The Modern Churchman*, 29(4), 32-36.
(Topics):theology.

Innes, K. (1987b). *Caring for the Earth: the environment, Christians, and the church.* Bramcote, Nottinghamshire: Grove Books Limited.
(Topics):ethics, book.

Institute for Ecumenical Research (1984). Consultation titled "Theology of creation -- contributions and deficits of our confessional traditions." *The Ecumenical Review*, 35(April).
(Topics):theology.

Institute for Ecumenical Research (1985). Tensions in contemporary theology of creation: an ecumenical challenge. *The Ecumenical Review*, 37, 360-370.
(Topics):popular.

International Consultation on Simple Lifestyle (1980). *An evangelical commitment to simple lifestyle.* Lausanne Committee on World Evangelization's Lausanne Theology and Education Group and the World Evangelical Fellowship's Theological Commission's Unit on Ethics and Society.
For copies, write to Unit on Ethics and Society, World Evangelical Fellowship, 300 W. Apsley St., Philadelphia, PA 10144.
(Topics):lifestyle.

Irish, E. R. (1980a). Benefits of nuclear power outweigh its risks. *Journal of the American Scientific Affiliation*, 32(2), 92-96.
(Topics):nuclear.

Irish, E. R. (1980b). Perspective on energy technology choices. *Journal of the American Scientific Affiliation*, 32(2), 112-114.

(Topics):nuclear.

Isenhart, C. (1987). Thompson farm: productive, profitable, regenerative. *Earthkeeping: A Quarterly on Faith and Agriculture*, 3(5), 17-18.
(Topics):land.

Isenhart, C. (1988, Sept. 11). Christians & ecologists meet on environment. *National Catholic Register*, pp. 1+.
(Topics):popular.

Jackson, B., & Dubos, R. (1972). *Only one earth*. London: Penguin Books.
(Topics):book.

Jackson, W. (1987a). Altars of unhewn stone. In F. W. Krueger (Ed.), *Christian ecology: building an environmental ethic for the twenty-first century* (pp. 38-41). North Webster, Indiana: The North American Conference on Christianity and Ecology.
(Topics):agriculture, land.

Jackson, W. (1987b). *Altars of unhewn stone: science and the Earth*. San Francisco: North Point Press.
(Topics):land.

Jackson, W. (1987c). Statement by Dr. Wes Jackson at press conference. In F. W. Krueger (Ed.), *Christian ecology: building an environmental ethic for the twenty-first century* (p. 7). North Webster, Indiana: The North American Conference on Christianity and Ecology.
(Topics):popular.

Jackson, W., Berry, W., & Colman, B. (1984). *Meeting the expectations of the land: essays in sustainable agriculture and stewardship*. San Francisco: North Point Press.
(Topics):land.

Jacobson, B. (1985). The mystery of creation. *Earthkeeping: A Quarterly on Faith and Agriculture*, 1(1), 17-18.

(Topics):popular.

Jaki, S. L. (1978). *The road of science and the ways to God.* Chicago: University of Chicago Press.
(Topics):book.

Jaoudi, M. (1987). Prayer and the sacredness of the Earth. In F. W. Krueger (Ed.), *Christian ecology: building an environmental ethic for the twenty-first century* (p. 24). North Webster, Indiana: The North American Conference on Christianity and Ecology.
(Topics):popular.

Jappe, F. (1978). A journal symposium -- the recombinant DNA controversy: a Christian perspective favoring recombinant DNA research. *Journal of the American Scientific Affiliation, 30*(2), 78.
(Topics):genetic engineering.

Jegen, M. E. (1987a). The church's role in healing the Earth. In W. Granberg-Michaelson (Ed.), *Tending the garden: essays on the gospel and the Earth* (pp. 93-113). Grand Rapids: Eerdmans Publishing Company.
(Topics):popular.

Jegen, M. E. (1987b). The local church's role in Earth healing. In F. W. Krueger (Ed.), *Christian ecology: building an environmental ethic for the twenty-first century* (p. 96). North Webster, Indiana: The North American Conference on Christianity and Ecology.
(Topics):popular.

Jegen, M. E. (1989a). Calling us to our senses: the church's role in healing the Earth. *Reformed Journal, 39*(10), 13-23.
    The article was previously published -- Jegen 1987a.
(Topics):popular.

Jegen, M. E. (1990). An encounter with God: keeping the Sabbath for the sake of Creation. *Sojourners, 19*(F/Mr), 15-17.

(Topics):popular.

Jegen, M. E., & Manno, B. V. (Eds.) (1978). *The Earth is the Lord's: essays on stewardship*. New York: Paulist Press.
(Topics):book, stewardship.

Jensen, F. L., & Tilberg, C. W. (1972). *The human crisis in ecology*. New York: Board of Social Ministry, Lutheran Church in America.
(Topics):book.

Jobling, D. (1977). And have dominion: the interpretation of Genesis 1:28 in Philo Judaeus. *Journal for the Study of Judaism in the Persian, Hellenistic and Roman Period*, 8(January), 50-82.
(Topics):Judaism, theology.

John Paul II, Pope. (1990). Peace with God the creator: peace with all of Creation. *Earth Ethics*, 1(3), 11.
(Topics):ethics.

Johnson, R. A. (1978). Brethren and the stewardship of energy. *Brethren Life and Thought*, 23, 199-209.
(Topics):energy.

Johnston, R. K. (1987). Wisdom literature and its contribution to a biblical environmental ethic. In W. Granberg-Michaelson (Ed.), *Tending the garden: essays on the gospel and the earth* (pp. 66-82). Grand Rapids: Eerdmans Publishing Company.
(Topics):theology.

Jones, D. L. (1971). Corporateness and the ecological conscience. *Religion in Life*, 40, 203-209.
(Topics):popular.

Jones, G. D. (1978). A journal symposium -- the recombinant DNA controversy: dangerous territory, not forbidden knowledge. *Journal of the American Scientific Affiliation*, 30(2), 79-80.

(Topics):genetic engineering.

Joranson, P. N. (1977). Faith-man-nature group and a religious environmental ethic. *Zygon*, 12(2), 175-179.
(Topics):popular.

Joranson, P. N., & Anderson, A. C. (Eds.) (1973). *Religious reconstruction for the environmental future.* South Coventry, Conn.: Faith-Man-Nature Group (proceedings of conference).
(Topics):book.

Joranson, P. N., & Butigan, K. (Eds.) (1985). *Cry of the environment: rebuilding the Christian creation tradition.* Santa Fe: Bear & Company.
(Topics):book.

Journal of the American Scientific Affiliation (editorial) (1973). Man has a positive responsibility to manage nature. *Journal of the American Scientific Affiliation*, 25, 3-4.
(Topics):popular.

Journal of the American Scientific Affiliation (thematic issue) (1969). *Journal of the American Scientific Affiliation*, 21(2), 33-47.
　　　Thematic issue dealing with popular environmental issues.
(Topics):popular.

Journal of the American Scientific Affiliation (thematic issue) (1974). *Journal of the American Scientific Affiliation*, 26(1), 1-21.
　　　Thematic issue dealing with the population crisis.
(Topics):population.

Journal of the American Scientific Affiliation (thematic issue) (1978). *Journal of the American Scientific Affiliation*, 30(2), 73-81.
　　　Thematic issue on recombinant DNA Research.
(Topics):genetic engineering.

Kabilsingh, C. (1987). How Buddhism Can Help Protect Nature. In S. Davies (Ed.), *Tree of Life: Buddhism and Protection of Nature with a Declaration on Environmental Ethics from His Holiness the Dalai Lama and an Introduction by Sir Peter Scott* (pp. 7-16). Buddhist Perception of Nature.
    (Topics):Buddhism.

Kahl, B. (1987). Human culture and the integrity of creation: biblical reflections on Genesis 1-11. *Ecumenical Review*, 39, 128-137.
    (Topics):theology.

Kanten, A. (1986). The erosion of soil and culture. *Word & World*, 6(1), 35-39.
    (Topics):land.

Kantonen, T. A. (1956). *A theology for Christian stewardship*. Philadelphia: Muhlenberg Press.
    (Topics):stewardship, theology.

Kantzer, K. S. (Ed.) (1987). *Applying the scriptures: papers from ICBI summit III*. Grand Rapids, MI: Academie Books, Zondervan Publishing House.
    See chapter 16.
    (Topics):theology, ethics.

Kantzer, K. S. (1989). The imperatives of wealth: the Christian ideal. *Christianity Today*, 33(8), 39-40.
    (Topics):lifestyle.

Kardong, T. (1983). Monks and the land. *Cistercian Studies*, 18(2), 135-148.
    (Topics):land.

Karff, S. E. (1974). Man's power and limits in a technological age. *Judaism*, 23, 161-173.
    (Topics):Judaism.

Kaufman, G. (1972). A problem for theology: the concept of nature. *Harvard Theological Review*, 65, 337-366.
(Topics):theology.

Kay, J. (1988). Concepts of nature in the Hebrew Bible. *Environmental Ethics*, 10(4), 309-327.
(Topics):ethics.

Keenan, B. (1973). Energy crisis and its meaning for American culture. *The Christian Century*, 90(27), 756-759.
(Topics):energy.

Keener, C. (1982). Toward a holistic view of nature: a process perspective. In E. R. Squiers (Ed.), *The environmental crisis: the ethical dilemma* (pp. 9-20). Mancelona, MI: Au Sable Institute of Environmental Studies.
(Topics):ethics.

Keller, J. A. (1971). Types of motives for ecological concern. *Zygon*, 6, 197-209.
(Topics):popular.

Kellert, S. R., & Berry, J. K. (1980). *Phase III: knowledge, affection and basic attitudes toward animals in American society.* Part of a U.S. Fish and Wildlife Service funded study to determine knowledge and attitudes about the environment. A question measuring frequency of attendance at religious services permitted correlation between attitude and knowledge vs. frequency of attendance.
(Topics):popular, knowledge.

Kelly, T. (1990). Wholeness: ecological and catholic. *Pacifica*, 3(2):201-223.
(Topics):popular.

Kennedy, E. J. (1970). Take care of the Earth. *Eternity*, 21(5), 15-16.
(Topics):popular.

Kennedy, E. J. (1973). The Christian and ecology. *Journal of the American Scientific Affiliation*, 25(1), 1-4.
(Topics):popular.

Kenyon, J. (1989, January/March). How shall we care for the Earth? *Together: A Publication of World Vision*, pp. 1-2.
(Topics):popular.

Kerr, H. T. (1972). Ecosystems and systematics. *Theology Today*, 29(1), 104-119.
(Topics):theology.

Khalil, I. J. (1978). Ecological crisis: an Eastern Christian perspective. *Saint Vladimir's Theological Quarterly*, 4, 193-211.
Reprinted in *Epiphany*, 6(3):38-50.
(Topics):theology.

Khalil, I. J. (1987). An Eastern Christian view of ecology. In F. W. Krueger (Ed.), *Christian ecology: building an environmental ethic for the twenty-first century* (p. 27). North Webster, Indiana: The North American Conference on Christianity and Ecology.
(Topics):Eastern Orthodox.

Kim, Y. B. (1979). The sustainable society: an Asian perspective. *Ecumenical Review*, 31(April), 169-178.
(Topics):popular, technology.

Kimbrell, A. (1987). The churches and genetic engineering. In F. W. Krueger (Ed.), *Christian ecology: building an environmental ethic for the twenty-first century* (p. 93). North Webster, Indiana: The North American Conference on Christianity and Ecology.
(Topics):genetic engineering.

Kime, R. (1983). Earth and stewardship: a conversation with John Todd. *Epiphany*, 3(4), 38-46.
(Topics):popular.

Kime, R., Shaw, M., & Garrett, A. (1983). A garden in Eden: considerations for a rural community. *Epiphany*, 3(4), 47-57.
(Topics):popular.

King, J. (1987). Neighbor networking: a united church response to the rural crisis in Canada. *Earthkeeping: A Quarterly on Faith and Agriculture*, 3(4), 10-13.
(Topics):land.

King, P. (1977). Global stewardship. In B. Kaye (Ed.), *The changing world* Glasgow: Collins.
(Topics):book, stewardship.

King, R. H. (1974). Ecological motif in the theology of H. Richard Niebuhr. *Journal of the American Academy of Religion*, 42, 339-343.
(Topics):theology.

King, R. J., & Piltz, R. (1988, March). How much longer can nature sustain us? *Christian Social Action*, pp. 18-21.
(Topics):pollution, agriculture.

Kingston, A. R. (1968). Christian duty and animal welfare. *Theology*, 71, 250-256.
(Topics):animal rights, ethics.

Kirby, W. (1835). *On the power, wisdom and goodness of God as manifested in the creation of animals and in their history, habits, and instincts.* London: W. Pickering.
(Topics):book.

Kirkland, L. A. (1987). Church activism to stop toxic pollution. In F. W. Krueger (Ed.), *Christian ecology: building an environmental ethic for the twenty-first century* (pp. 105-106). North Webster, Indiana: The North American Conference on Christianity and Ecology.
(Topics):pollution, toxic waste.

Klay, R. K. (1986). *Counting the cost: the economics of Christian stewardship*. Grand Rapids: Wm. B. Eerdmans Pub. Co.
(Topics):economics, stewardship

Klewin, T. (1975), The Christian and ecology: consumption ethic must be replaced. *Our Sunday Visitor Magazine*, pp. 1+.
(Topics):lifestyle.

Kline, D. (1987). The lessons of Amish agriculture. In F. W. Krueger (Ed.), *Christian ecology: building an environmental ethic for the twenty-first century* (pp. 41-43). North Webster, Indiana: The North American Conference on Christianity and Ecology.
(Topics):land, agriculture.

Klink, W. H. (1974). Environmental concerns and the need for a new image of man. *Zygon*, 9(December), 300-310.
(Topics):technology.

Klostermaier, K. K. (1973). World religions and the ecological crisis. *Religion*, 3, 132-145.
(Topics):popular.

Klotz, J. W. (1971a). Creationism and our ecological crisis. *Creation Research Society Quarterly*, 8, 15, 49.
One of the few references to stewardship of the creation to be found in the "creationist" literature.
(Topics):popular.

Klotz, J. W. (1971b). *Ecology crisis: God's creation and man's pollution*. St. Louis, Mo.: Concordia Publishing House.
(Topics):pollution, popular book.

Klotz, J. W. (1972). A reply by Dr. Klotz. *Creation Research Society Quarterly*, 8(4), 284-287.

Response to Robbins 1972 who argues that there is no ecological crisis and no need for the church to be concerned with caring for creation.
(Topics):popular.

Klotz, J. W. (1984). Creationist Environmental Ethic. *Creation Research Society Quarterly,* 21, 6-8.
Makes the point that all "creationists" should be creation stewards.
(Topics):popular.

Klotz, J. W. (1987). A response to stewardship of the environment. In K. S. Kantzer (Ed.), *Applying the scriptures: papers from ICBI summit III* (pp. 489-496). Grand Rapids, MI: Academie Books, Zondervan Publishing House.
(Topics):theology.

Knight, P. (1967). *The politics of conservation: Christians and the good Earth.* Alexandria, Va: The Faith-Man-Nature Group.
(Topics):popular book.

Knoch, M. (1990). "Dadiva e Conquista" (gift and conquest). *International Review of Mission,* 79(314), 167-177.
Primary focus is on just distribution of land in the LDCs.
(Topics):popular.

Knudsen, R. D. (1962). Ethics and birth control. *Journal of the American Scientific Affiliation,* 14(1), 7-11.
(Topics):ethics, population.

Koch, K. (1979). The Old Testament view of nature. *Anticipation (World Council of Churches),* 50.
(Topics):theology.

Kolk, A. (1989). Chewing the convention cud. *Earthkeeping: A Quarterly on Faith and Agriculture,* 5(1), 22-23.
Reflections on the 1988 Annual Convention of the Christian Farmers Federation of Alberta.

(Topics):land, agriculture.

Kolkman, J. (1986). Pesticides and poverty: a deadly combination. *Earthkeeping: A Quarterly on Faith and Agriculture*, 2(3), 8-10.
(Topics):pollution, pesticides.

Kolkman, J. (1987). Biotechnology: brave new world for agriculture. *Earthkeeping: A Quarterly on Faith and Agriculture*, 3(1), 11-15.
(Topics):genetic engineering, technology, agriculture.

Kolkman, J. (1988). Meeting the animal rights challenge. *Earthkeeping: A Quarterly on Faith and Agriculture*, 4(1), 4-7.
(Topics):animal rights.

Konvitz, M. R. (1972). *Judaism and human rights*. New York: W. W. Norton & Company, Inc.
See especially Part VI, "The Earth is the Lord's."
(Topics):Judaism.

Kottukapally, J. (1974). Nature and grace: a new dimension. *Thought*, 49(June), 117-133.
(Topics):theology.

Kraps, J. M. (1986). A call to agricultural conversion. *Sojourners*, 15(9), 12.
Report of the first conference meeting of "The Institute in Church and Rural Life Issues." A major theme of the conference was the proper response of the Church to the damage of the Creation by poor farming practices.
(Topics):land, agriculture.

Krause, G. (1982). Does matter matter. *Crux: A Quarterly Journal of Christian Thought and Opinion*, 18(June), 2-5.
(Topics):popular.

Kroeker, W. (1989). Harnessing the ox in Tanzania. *Earthkeeping: A Quarterly on Faith and Agriculture*, 5(2), 9-11.
(Topics):appropriate technology, development.

Kroll, L. C. (1982). Technology: Modern Ebal and Gerizem? A Chemist's Perspective. In E. R. Squiers (Ed.), *The environmental crisis: the ethical dilemma* (pp. 177-186). Mancelona, MI: Au Sable Institute of Environmental Studies.
(Topics):technology.

Krueger, F. (1985a). The eleventh commandment and the environmental crisis. *Epiphany*, 6(1), 16-24.
(Topics):popular.

Krueger, F. (1985b). To Heal the Earth. *Epiphany*, 6(1), 66-73.
(Topics):popular.

Krueger, F. (1987a). The history of the NACCE. In F. W. Krueger (Ed.), *Christian Ecology: building an environmental ethic for the twenty-first century* (pp. 2-4). North Webster, Indiana: The North American Conference on Christianity and Ecology.
(Topics):popular.

Krueger, F. (1987b). Wilderness retreats: part three: experiencing the numinous in nature. In F. W. Krueger (Ed.), *Christian ecology: building an environmental ethic for the twenty-first century* (p. 57). North Webster, Indiana: The North American Conference on Christianity and Ecology.
(Topics):wilderness, popular.

Krueger, F. (Ed.) (1988). *Christian ecology: building an environmental ethic for the twenty-first century*. San Francisco: The North American Conference on Christianity and Ecology.
(Topics):popular book.

Krueger, R. R. (1971). Towards an environmental ethic. *Crux*, 8(2), 27-32.
   (Topics):popular.

Kuhn, H. B. (1970). Environmental stewardship. *Christianity Today*, 14(16), 758-759.
   (Topics):popular, stewardship.

Kuhns, W. (1969). *Environmental man*. New York: Harper & Row, Publishers.
   (Topics):popular book.

LaBar, M. (1974). Message to polluters from the Bible. *Christianity Today*, 18(21), 1186-1190.
   (Topics):pollution, popular.

Lamm, N. (1964-65). Man's position in the universe: a comparative study of the views of Saadia Gaon and Maimonides. *The Jewish Quarterly Review*, 55(3), 208-234.
   (Topics):Judaism, theology.

Lamm, N. (1971). *Faith and doubt: studies in traditional Jewish thought*. New York: Ktav Publishing House, Inc.
   (Topics):Judaism.

Lamm, N. (1972). The Earth is the Lord's. In M. R. Konvitz (Ed.), *Judaism and human rights*. New York: W. W. Norton.
   (Topics):Judaism.

Lampe, G. W. H. (1964). The New Testament doctrine of Ktisis. *Scottish Journal of Theology*, 17(4), 449-462.
   Reported by Santmire 1985 to be anthropocentric.
   (Topics):theology.

Landes, G. M. (1978). Creation and liberation. *Union Seminary Quarterly Review*, 33(2), 79-89.
   (Topics):theology.

Landon, C. H. (1990). *Sacred cosmos: the implications of ancient pagan traditions for modern theology of nature.* D.Min., Meadville/Lombard Theological School.
(Topics):theology.

Lawton, K. A. (1990). Is there room for prolife environmentalists? *Christianity Today,* 34(13), 46-47.
(Topics):population.

Learner, M. (Ed.) (1990). Special feature: Earth Day 1990. *Tikkum: A Bimonthly Jewish Critique of Politics, Culture & Society,* 5(Mr/Ap), 48-81.
Thematic issue for Earth Day 1990. Two of the papers in this special issue of *Tikkum* examine environmental issues from a Jewish perspective: Cohen 1990 and Waskow 1990
(Topics):Judaism.

Le-Chau, L. (1989, January-March). Louga: 'I will plant in the desert.' *Together: A Publication of World Vision,* p. 10.
(Topics):popular, land.

Lechte, R. E. (1990). Partnerships for ecological well-being. *The Ecumenical Review,* 42(April), 157-161.
Focuses on environmental organizations.
(Topics):popular.

Lee, C. (1990). The integrity of justice: evidence of environmental racism. *Sojourners,* 19(F/Mr), 20-25.
The focus of the article is the disproportionate impact of pollution on and location of toxic waste sites in racial minority communities.
(Topics):ecojustice, racial justice, pollution, toxic waste.

Leeuwen, R. C. V. (1991). Resurrection and the vindication of Creation. In C. B. DeWitt (Ed.), *The environment and the Christian: what does the New Testament say about the environment?* Grand Rapids, Michigan: Baker Book House.
(Topics):theology.

Leidke, G. (1980). Solidarity in conflict. In R. L. Shinn (Ed.), *Faith and science in an unjust world: report of the World Council of Churches' conference on faith, science and the future, Vol. 1: plenary presentations.* Philadelphia: Fortress Press.
(Topics):popular.

Leiss, W. (1972). *The domination of nature.* New York: George Brazileer.
See references to Christianity in this basically secular work.
(Topics):history.

Leliaert, R. (1970). All things are yours. *Homiletic and Pastoral Review*, 70, 573-578+.
(Topics):popular.

Leopold, A. (1974). *A Sand County Almanac.* New York: Sierra Club/Ballantine.
A most important secular work describing some of the first steps of ecological restoration.
(Topics):popular.

Lepkowski, H., & Lepkowski, W. (1987). Opportunities and obstacles in relating Christian salvation belief to ecological renewal. In F. W. Krueger (Ed.), *Christian ecology: building an environmental ethic for the twenty-first century* (p. 103). North Webster, Indiana: The North American Conference on Christianity and Ecology.
(Topics):popular.

Le Roux, P. J. (Ed.) (1987). *Environment conservation: why and how?* Pretoria, South Africa: University of South Africa.
(Topics):conservation, resources.

Lewthwaite, G. R. (1974). Moral restraint rather than coercion. *Journal of the American Scientific Affiliation*, 26(1), 18.
(Topics):population.

Lieder, D. L. (1984) *Marginality: a theological indictment: referring to the problem of the human relationship to the natural world.* D.Min., School of Theology at Claremont.
(Topics):theology.

Lilburne, G. R. (1989). *A sense of place: a Christian theology of the land.* Nashville: Abingdon Press.
(Topics):theology, land.

Lill, C. (1989). This land is home to me. *The Egg: A Journal of Eco-Justice,* 9(1), 11,16.
(Topics):popular.

Limburg, J. (1971). What does it mean to 'have dominion over the Earth'? *Dialog,* 10(Summer), 221-223.
(Topics):popular.

Limburg, J. (1981). *A Good Land.* Minneapolis: Augsburg.
(Topics):land, book.

Limouris, G. (1989). Integrity of Creation and Earth. *Mid-Stream,* 28(3), 249-262.
From the Greek Orthodox perspective.
(Topics):theology, popular.

Lindsell, H. (1976). Lord's day and natural resources. *Christianity Today,* 20(16), 8-12.
(Topics):resources, popular.

Link, C. (1984). Ecological crisis and Christian ethics. *Theology Digest,* 31(2), 149-153.
(Topics):ethics.

Linn, D. W. (1973). Christian -- it's your environment too. *Journal of the American Scientific Affiliation,* 25(1), 13-16.
(Topics):popular.

Linn, G. (1990). From section III in San Antonio 1989 to section I in Canberra 1991. *International Review of Mission*, 79(314), 195-205.
> Report on the World Council of Churches environmental work.
> (Topics):popular.

Linzey, A. (1976). *Animal rights: a Christian assessment of man's treatment of animals.* London: SCM Press.
> (Topics):animal rights, ethics.

Linzey, A. (1987). *Christianity and the rights of animals.* New York: Crossroad Publishing Co.
> (Topics):animal rights.

Linzey, A. (1990). Why Christians should care. *Christianity Today*, 34(9), 22.
> (Topics):animal rights.

Little, J. A. (1984) *Esse/essence and grace: a theological inquiry into Thomist methodology (creation, covenant, christology).* Ph.D., Marquette University.
> (Topics):theology.

Livingston, J. C. (1971). The ecological challenge to Christian ethics. *Christian Century*, 88(48), 140-141.
> (Topics):ethics, popular.

Loader, J. A. (1987). Image and order: Old Testament perspectives on the ecological crisis. In W. S. Vorster (Ed.), *Are We Killing God's Earth?* (pp. 6-28). Pretoria, South Africa: University of South Africa.
> (Topics):theology.

Lofgren, R. (1987). Sharing the stewardship of our spaceship earth. In L. Kenworthy (Ed.), *Friends face the world: continuing and current Quaker concerns* (pp. 136-146). Kennett Square, PA: Quaker Publications.
> (Topics):resources, stewardship.

Loftus, M. F. (1989). The earth is the Lord's: reflection on a sub-theme of the 1989 WCC conference on mission and evangelism. *SEDOS Bulletin*, 6, 208-212.
(Topics):popular.

Logan, W. M. (1957). *In the beginning God: the meaning of Genesis 1-11*. Madison, Wis.: Adult Christian Education Foundation.
(Topics):popular.

Lohman, C. (1987). A theo-ecology of diet. In F. W. Krueger (Ed.), *Christian ecology: building an environmental ethic for the twenty-first century* (pp. 35-36). North Webster, Indiana: The North American Conference on Christianity and Ecology.
(Topics):popular.

Longacre, D. J. (1976). *More-with-less cookbook*. Scottdale, PA: Herald Press.
(Topics):lifestyle.

Longacre, D. J. (1980). *Living more with less*. Scottdale, PA: Herald Press.
(Topics):lifestyle.

Longwood, M. (1972). Toward an environmental ethic. In G. Devine (Ed.), *That they may live; theological reflections of the College Theological Society* (pp. 47-68). St. Paul, Minn.: Alba House.
(Topics):ethics.

Longwood, M. (1973). Common good: an ethical framework for evaluating environmental issues. *Theological Studies*, 34, 468-480.
(Topics):ethics.

Lønning, P. (1984). Theology of creation -- contributions and deficits of our confessional traditions. *The Ecumenical Review*, 36, 204-213.
(Topics):theology.

Lønning, P. (1985). Tensions in contemporary theology of creation: an ecumenical challenge? *The Ecumenical Review*, 37, 360-370.
(Topics):theology.

Lowdermilk, W. C. (1933). The eleventh commandment. *American Forests*, 46(January):12-15.
(Topics):popular.

Loy, R. P. (1982). Politics and the environment: toward a new public philosophy. In E. R. Squiers (Eds.), *The environmental crisis: the ethical dilemma* (pp. 209-226). Mancelona, MI: Au Sable Institute of Environmental Studies.
(Topics):ethics.

Lutheran Quarterly (thematic issue) (1971). Reflections on theological symbols, man and nature. *Lutheran Quarterly*, 23(4), 303-387.
(Topics):popular.

Lutheran World Federation (7th Assembly) (1985). *Statements by the 7th Assembly*, No. 19-20. Lutheran World Federation.
See pages 175-187.
(Topics):popular.

Lutz, C. P. (Ed.) (1980). *Farming the Lord's land: Christian perspectives on American agriculture.* Minneapolis: Augsburg Publishing House.
(Topics):land.

Lutz, P. E. (1975a). *The environmental crisis: business as usual?* New York: Division of Theological Studies, Lutheran Council in the U.S.A.
(Topics):popular.

Lutz, P. E. (1975b). Quality of life--which way Christians? *Dial*, 14(Spring), 144-147.

(Topics):popular.

Lutz, P. E. (1987). Interdependence: a central principle for ecology and theology. In F. W. Krueger (Ed.), *Christian ecology: building an environmental ethic for the twenty-first century* (pp. 59-60). North Webster, Indiana: The North American Conference on Christianity and Ecology.
(Topics):Gaia Hypothesis, popular.

Lutz, P. E., & Santmire, H. P. (1972). *Ecological renewal.* Philadelphia: Confrontation Books.
(Topics):book.

Lynch, T. B. (1990). Two worlds join to 'preserve the earth.' *Christianity and Crisis*, 50(7):142-144.
Report on the call from scientists (see Sagan 1990) for the religious community to join in preserving and cherishing the earth.
(Topics):popular.

Maahs, K. H. (1972) *The theology of human ecological responsibility in the Old Testament.* Ph. D., Southern Baptist Theological Seminary.
(Topics):theology.

Maatman, R. (1974). First things first. *Journal of the American Scientific Affiliation*, 26(1), 18-19.
(Topics):population.

Macer, D. (1990). Genetic engineering in 1990. *Science and Christian Belief*, 2(1), 25-40.
(Topics):genetic engineering.

MacGillis, M. T. (1987a). Genesis Farm: a learning center for re-inhabiting the Earth. In F. W. Krueger (Ed.), *Christian ecology: building an environmental ethic for the twenty-first century* (p. 69). North Webster, Indiana: The North American Conference on Christianity and Ecology.
(Topics):land, agriculture.

MacGillis, M. T. (1987b). Poverty, chastity and obedience: opening the meaning of the ecological age. In F. W. Krueger (Ed.), *Christian ecology: building an environmental ethic for the twenty-first century* (p. 18). North Webster, Indiana: The North American Conference on Christianity and Ecology.
(Topics):popular.

MacKay, A. I. (1950). *Farming and gardening in the Bible.* Emmaus, PA: Rodale Press.
(Topics):land, agriculture.

Macquarrie, J. (1971). Creation and environment. *The Expository Times*, 83(October), 4-9.
Reprinted in Spring and Spring, eds. 1974.
(Topics):theology, history.

Macquarrie, J. (1975). The idea of a theology of nature. *Union Seminary Quarterly Review*, 30(2/4), 69-75.
(Topics):theology.

Malloch, T. R. (1982). U. S. energy policy: past, present, and future. In E. R. Squiers (Eds.), *The environmental crisis: the ethical dilemma* (pp. 261-278). Mancelona, MI: Au Sable Institute of Environmental Studies.
(Topics):energy.

Malone, N. M. (Ed.) (1989). Repurposing education: the American college in the ecological age. *Religion and Intellectual Life*, 6, 7-69.
(Topics):popular.

Maloney, G. A. (1968). *The cosmic Christ.* New York: Sheed and Ward.
(Topics):popular book, theology.

Manahan, R. E. (1982). *A re-examination of the cultural mandate: an analysis and evaluation of the dominion materials.* Th.D., Grace Theological Seminary.
(Topics):theology.

Manahan, R. E. (1991). Christ as second Adam: where the first Adam fails, the second triumphs. In C. B. DeWitt (Ed.), *The environment and the Christian: what does the New Testament say about the environment?* Grand Rapids, Michigan: Baker Book House.
(Topics):theology.

Mann, P. (1986). Justice in the ecological age. *Studies in Formative Spirituality*, 7, 349-359.
(Topics):creation spirituality.

Marberry, T. L. (1982) *The place of the natural world in the theology of the Apostle Paul.* Ph.D., Baylor University.
(Topics):theology.

Marietta, D. E. (1977). Religious models and ecological decision making. *Zygon*, 12(June), 151-166.
(Topics):popular.

Markus, J., De Boer, J. De Boer, K. Schlegel, C., Vanden-Berg, B., and Oegema, T. (1991). Caring for Creation: is stewardship enough? *Earthkeeping: a quarterly on faith and agriculture*, 6(3):6-10.
(Topics):agriculture, land.

Marshall, K. (1983). The Shaker community. *Epiphany*, 3(3), 88-93.
(Topics):lifestyle.

Marshall, P. (1984). Is technology out of control? *Crux*, 20(3), 3-9.
(Topics):technology.

Marshall, P. (1985). A Christian view of economics. *Crux*, 21(1), 3-6.
(Topics):stewardship.

Martensen, D. F. (1970a). Concerning the ecological matrix of theology. *Zygon*, 5(4), 353-369.

(Topics):theology.

Martensen, D. F. (1970b). Theologizing in an ecological matrix. *Dialog*, 9(Summer), 192-199.
(Topics):theology.

Martensen, D. F. (1971). Editorial commentary: religious implications of the ecological crisis. *The Lutheran Quarterly*, 23(4), 303-304.
(Topics):popular.

Martin, K. A. (1980). Biblical mandates and the human condition. *Journal of the American Scientific Affiliation*, 32(2), 74-77.
(Topics):nuclear.

Masse, B. (1970). Christians and the crisis of environment. *America*, 123, 82.
(Topics):popular.

Massey, M. (1985). *Defense of the peaceable kingdom*. Religious Society of Friends, Oakland, California, Pacific Yearly Meeting.
(Topics):book.

Massey, M. (1987). The swords at Eden's gate: part 1: orientation; the ozone layer. In F. W. Krueger (Ed.), *Christian ecology: building an environmental ethic for the twenty-first century* (pp. 58-59). North Webster, Indiana: The North American Conference on Christianity and Ecology.
(Topics):air pollution, ozone.

Matthews, A. D. (1965). The prophetic doctrine of creation. *Church Quarterly Review*, 166, 141-149.
(Topics):theology.

Mattson, L. (Ed.) (1985). *God's good Earth: Christian values in outdoor education*. Duluth, Minn.: Camping Guideposts.
(Topics):book.

Maxey, M. N. (1980). Nuclear waste: beyond Faust and fate. *Journal of the American Scientific Affiliation*, 32(2), 97-101.
(Topics):nuclear.

Mayers, M. (1972). Ecology in the Old Testament. *His*, 32(6), 14-16.
(Topics):popular.

Mayers, R. B. (1972). Transcendence and ecology. *Reformed Review*, 25(Spring), 154-159.
(Topics):popular.

Maynard-Reid, P. U. (1989). The community of wealth: called to share. *Christianity Today*, 33(8), 37-39.
(Topics):lifestyle.

McAllister, J. (1987). Creation-centered spirituality: healing ourselves, healing the Earth. In F. W. Krueger (Ed.), *Christian ecology: building an environmental ethic for the twenty-first century* (p. 26). North Webster, Indiana: The North American Conference on Christianity and Ecology.
(Topics):creation spirituality.

McCarthy, T. (1988, Sept. 16). Christian ecologists chart return trip to Eden. *National Catholic Reporter*, pp. 1+.
(Topics):popular.

McClintock, H. (1971). The think-tank approach: one way of beginning education in ecology. *Religious Education*, 66(1), 55-61.
(Topics):popular.

McCormack, A. (1972). Subduing the Earth. *Frontier*, 15, 35-38.
(Topics):ethics.

McCormick, R. A. (1971). Toward an Ethics of Ecology. *Theological Studies*, 32(March), 97-107.
(Topics):theology.

McCoy, J. D. (1985). Towards a theology of nature. *Encounter, 46*, 213-228.
(Topics):theology.

McCrary, D. (1987). Clearing the air. *Eternity, 38*(7/8), 4. Editorial.
(Topics):popular.

McDaniel, J. (1986). Christian spirituality as openness to fellow creatures. *Environmental Ethics, 8*(1), 33-46.
(Topics):ethics, theology.

McDaniel, J. (1987). Christianity and the pursuit of wealth. *Anglican Theological Review, 69*, 349-361.
(Topics):theology, economics, lifestyle.

McDaniel, J. (1988). Land ethics, animal rights, and process theology. *Process Studies, 17*, 88-102.
(Topics):animal rights, land, ethics, theology.

McDaniel, J. (1989). *Earth, sky, gods and mortals.* Mystic, Conn.: Twenty-Third Publications.
(Topics):popular book.

McDaniel, J. (1990). *Of God and pelicans.* Philadelphia: Westminster/John Knox Press.
(Topics):popular book.

McDaniel, T. F. (1987). The right to work -- a gift from the creator. *Minister: A Journal of the American Baptist Ministers Council Speaking to the Practice of Ministry, 7*(2), 1-3.
    Provides a review of the Hebraic meaning of the key words "dominion," "subdue," "till," and "keep" in Genesis 1:26-29 and 2:15.
(Topics):theology.

McDonagh, E. (1986). *Between chaos and new creation: doing theology at the fringe.* Wilmington, DE: Michael Glazier, Inc.

(Topics):theology.

McDonagh, S. (1984). The good Earth: reflections on ecology and theology. *Furrow*, 35, 430-440.
(Topics):land.

McDonagh, S. (1986). *To care for the Earth: a call to a new theology*. London: Chapman.
(Topics):land, theology.

McDonagh, S. (1988). International debt and ecology. *Doctrine and Life*, 38, 508-517.
(Topics):economics, popular.

McDonald, J. H. (1984). *Stewardship of creation: some implications for economic theory and policy*. Ph.D., Louisiana State University.
(Topics):economics, stewardship.

McFadden, T. M. (1963). *The relationship between nature and grace: a survey of 20th century theological opinion*. S.T.D., The Catholic University of America.
(Topics):theology.

McFague, S. (1987). *Models of God: theology for an ecological/nuclear age*. Philadelphia: Fortress Press.
(Topics):popular, nuclear.

McFague, S. (1989). The theologian as Advocate. *Theological Education*, 25, 79-97.
(Topics):theology.

McFarland, J. R. (1988, September/October). Preaching through the agricultural crisis. *The Christian Ministry*, pp. 13-14.
(Topics):agriculture, land.

McHarg, I. (1969). *Design with nature*. Garden City: Natural History Press.

Supports the argument that the Judeo-Christian community bears blame for the ecological crisis.
(Topics):book.

McHarg, I. (1970). Values, process, and form. In R. Dish (Ed.), *Ecological conscience* (pp. 21-36). Englewood Cliffs, N.J.: Prentice-Hall, Inc.
Supports the argument that the Judeo-Christian community bears blame for the ecological crisis.
(Topics):popular.

McIntyre, J. (1981). The theological dimension of the ecological problem. *The Scottish Journal of Religious Studies*, 2(Autumn), 83-96.
(Topics):theology.

McKeever, J. (1990). Father God and mother nature. *End-Times News Digest*, No. 141, 1-6.
A paper based on dispensational theology that is lacking in both its scientific understanding and Scriptural background.
(Topics):popular.

McKenzie, J. L. (1952). God and nature in the Old Testament. *Catholic Biblical Quarterly*, 14, 18-39; 124-145.
(Topics):history.

McKibben, B. (1990). The end of nature. *Earth Ethics: Evolving values for an Earth community*, 1(3), 1-8.
(Topics):ethics.

McPherson, J. (1982) *Ecological theology within the church and society programme of the World Council of Churches 1966-1979*. M.Litt., Durham University.
Copies at Durham, the WCC Library in Geneva, and St. Mark's Library in Canberra.
(Topics):theology.

McPherson, J. (1983). Ethics, theology, and the national conservation strategy. *St. Mark's Review*, 116, 17-23.

Focus is on Australia.
(Topics):ethics, theology.

McPherson, J. (1986). Towards an ecological theology. *The Expository Times*, 97(8), 236-240.
(Topics):theology.

McQuail, T. (1987). Banding together: the farming community vs. Ontario hydro. *Earthkeeping: A Quarterly on Faith and Agriculture*, 3(3), 5-7.
(Topics):land.

Mead, M. (1964). Introduction. In H. White (Ed.), *Christians in a technological era*. New York: Seabury Press.
(Topics):technology, popular.

Means, R. L. (1967, December 2). Why worry about nature? *Saturday Review*.
(Topics):popular.

Meed, M. (1990). Environment vs. salvation. *Faith Today*, 8(Mr/Ap), 34-36.
Discusses the lack of involvement of Canadian Christians in environmental matters and urges all Christians to recognize the care of Creation as part of their stewardship responsibilities.
(Topics):popular.

Meeker, J. W. (1988, Spring). Amplifications: continuing notes on the land ethic: the Assisi connection. *Wilderness*, p. 61-63.
(Topics):popular, St. Francis of Assisi.

Meeks, D. M. (1985). God and land. *Agriculture and Human Values*, 11(4), 16-27.
(Topics):land, theology.

Megivern, J. *God's good Earth and ours*. New York: The Christophers.

Date unknown, but probably published near 1970. (Topics):popular book.

Megivern, J. (1970). Ecology and the Bible. *Ecumenist*, 8, 69-71.
(Topics):popular.

Meland, B. E. (1974). Grace: a Dimension within Nature? *The Journal of Religion*, 54, 119-137.
(Topics):theology.

Mellert, R. (1973). Models and Metanoia. *Proceedings of the Catholic Philosophical Association*, 47, 142-152.
Suggests a move toward a more pantheistic view.
(Topics):popular.

Mertens, H. (1987). The doctrine of creation in ecological perspective. *Louvain Studies*, 12, 83-88.
(Topics):theology.

Meye, R. P. (1987). Invitation to wonder: toward a theology of nature. In W. Granberg-Michaelson (Ed.), *Tending the garden: essays on the gospel and the Earth* (p. 150). Grand Rapids, MI: Eerdmans Publishing Company.
(Topics):theology.

Meyer, A. (1987). War, conflict and environmental degradation: part one: overview. In F. W. Krueger (Ed.), *Christian ecology: building an environmental ethic for the twenty-first century* (pp. 72-74). North Webster, Indiana: The North American Conference on Christianity and Ecology.
(Topics):nuclear war.

Meyer, A., and Meyer, J. (1991). *Earth-keepers: environmental perspectives on hunger, poverty, and injustice.* Scottdale, PA.: Herald Press.
Foreword by Calvin B. DeWitt of Au Sable Institute.

(Topics):Biblical foundations, ethics, ecojustice, pollution, hazardous waste, agriculture, land, population, economics, genetic engineering, energy.

Meyer, L. (1984). Luther in the Garden of Eden: his commentary on Genesis 1-3. *Word & World*, 4(4), 430-436.
(Topics):theology.

Miles, S. J. (1985). The roots of the scientific revolution: reformed theology. *Journal of the American Scientific Affiliation*, 37(3), 158-168.
(Topics):history.

Milgrom, J. (1963). The biblical diet laws as an ethical system: food and faith. *Interpretation*, 17, 288-301.
(Topics):ethics.

Miller, D. E. (1979). The biblical basis for ecology. *Bretheren Life and Thought*, 24(Winter), 12-17.
(Topics):popular.

Miller, H. A. (1970). Man: created to be an Earth keeper. *Brethren Life and Thought*, 15(Spring), 74-78.
(Topics):popular.

Miller, V. (1988). Valuing animals in their own right. *Earthkeeping: A Quarterly on Faith and Agriculture*, 4(1), 8-10.
(Topics):animal rights.

Mische, P. M. (1978). Parenting in a hungry world. In M. E. Jegen & B. V. Manno (Eds.), *The Earth is the Lord's: essays on stewardship* (pp. 169-183). New York: Paulist Press.
(Topics):popular.

Mitcham, C., & Grote, J. (Eds.) (1984). Selected bibliography of theology and technology. In *Theology and technology: essays in Christian analysis and exegesis* (pp. 323-516). Lanham, MD: University Press of America.
(Topics):bibliography.

Mixter, R. L. (1970). Other literature. *Journal of the American Scientific Affiliation*, 22(2), 51-52.
(Topics):population.

Mixter, R. L. (1973). The population explosion. *Journal of the American Scientific Affiliation*, 25(1), 9-13.
(Topics):population.

Mixter, R. L. (1974). I am part of the problem. *Journal of the American Scientific Affiliation*, 26(1), 19.
(Topics):population.

Mixter, R. L. (1978). A journal symposium -- the recombinant DNA controversy: published pros and cons. *Journal of the American Scientific Affiliation*, 30(2), 75.
(Topics):genetic engineering.

Moberg, D. O. (1974). All economic and political systems inadequate. *Journal of the American Scientific Affiliation*, 26(1), 19-20.
(Topics):population.

Moberly, R. W. (1988). Did the serpent get it right? *The Journal of Theological Studies*, 39, 1-27.
(Topics):theology.

Modern Churchman (1970). Nature, Man and God. *Modern Churchman*, 14(1), 1-118.
Thematic issue.
(Topics):theology.

Moellering, R. L. (1971). Environmental crisis and Christian responsibility. *Concordia Journal*, 42, 176-182.
(Topics):popular.

Moerman, J. (1991). Farm finance policy ... and God: a reflective look at a Christian's experience on a government task force. *Earthkeeping: a quarterly on faith and agriculture*, 6(3):17-18

(Topics):land, agriculture.

Moldenke, H. N., & Moldenke, A. L. (1952). *Plants of the Bible*. Waltham, Mass.: Chronica Botanica Co. Reprinted in 1986 by Dover Publications.
(Topics):scriptural plants and animals.

Moltmann, J. (1972). Christian theology today. *New World Outlook*, 62, 483-490.
(Topics):theology.

Moltmann, J. (1979). *The future of creation*. Philadelphia: Fortress Press.
(Topics):theology.

Moltmann, J. (1985). *God in creation: a new theology of creation and the Spirit of God*. San Francisco: Harper & Row.
(Topics):theology.

Moltmann, J. (1986). The cosmic community: a new ecological concept of reality in science and religion. *Ching Feng: Quarterly Notes on Christianity and Chinese Religion and Culture*, 29(2-3), 93-105.
(Topics):theology.

Moltmann, J. (1988). The Ecological Crisis: Peace with Nature? *The Scottish Journal of Religious Studies*, 9, 5-18.
(Topics):theology.

Moncrief, L. (1970). The cultural basis for our environmental crisis. *Science*, 170, 508-512.
Response to White's 1967 paper; reprinted in Spring and Spring, eds. 1974.
(Topics):theology.

Monsma, S. V. (Ed.) (1986). *Responsible technology: a Christian perspective*. Grand Rapids, MI: Eerdmans.
Provides a distinctly Christian perspective on the impact and role of technology in our society. Produced by the

Calvin Center for Christian Scholarship. (Topics):technology.

Monsma, S. V. (1988, November/December). Caring for God's Earth: Public Policy and the Environment. *ESA Advocate*, p. 1A-4A. (Topics):popular.

Montefiore, H. (1965). Man's dominion. *Theology*, 68, 41-46. (Topics):popular.

Montefiore, H. (1966). Man's dominion. In E. Barker (Ed.), *The responsible church*. London: S.P.C.K., Holy Trinity Church. (Topics):book.

Montefiore, H. (1970). *Can Man Survive?* London: Collins. (Topics):book.

Montefiore, H. (1971). Ecology, theology, and posterity. *New Scientist and Science Journal*, 49, 316-318. (Topics):popular.

Montefiore, H. (Ed.) (1975). *Man and nature*. London: Collins. (Topics):theology.

Montefiore, H. (1977). Man and nature: a theological assessment. *Zygon*, 12(3), 199-211. (Topics):theology.

Montesano, C. (1984). A Catholic worker philosophy of the land. *Catholic Agitator*, 14(4), 7. (Topics):land.

Moody Monthly (editorial) (1970). Christians and Ecology. *Moody Monthly*, 71(9), 8. (Topics):popular.

Mooney, P. R. (1985). Impact on the farm: the other side of bio-technology. *Earthkeeping: A Quarterly on Faith and Agriculture*, 1(2), 4-9.
(Topics):genetic engineering.

Moore, J., & Arena, R. (1990). Feeling the squeeze: are we destroying God's Earth? *The American Baptist*, 188(1), 10-17.
(Topics):popular.

Moore, P. D. (1990). The exploitation of forests. *Science & Christian Belief*, 2(2), 131-140.
(Topics):resources, forests.

Moore, S. W., & Jappe, F. (1980). Christianity as an ethical matrix for no-grow economics. *Journal of the American Scientific Affiliation*, 32(3), 164-168.
(Topics):economics, ethics.

Morikawa, J. (1974). Theological meaning of ecojustice through institutional change. *Foundations: A Baptist Journal of History and Theology*, 17(2), 100-105.
(Topics):ecojustice.

Morikawa, J. (1975a). Evangelistic life style in an ecojust universe. *Foundations*, 18, 272-281.
(Topics):lifestyle, popular, ecojustice.

Morris, R. K., & Fox, M. W. (Eds.) (1978). *On the fifth day: animal rights and human ethics*. Washington D.C.: Acropolis Books.
(Topics):animal rights.

Morrison, W. D. (1988). Animal welfare in the Bible. *Earthkeeping: A Quarterly on Faith and Agriculture*, 4(1), 11-12.
(Topics):animal rights.

Morton, J. P. (1980). Listen to the Earth. *Christianity and Crisis: A Christian Journal of Opinion*, 40(Feb. 4), 10-12.

(Topics):popular.

Morud, J. (1989a). Caretaking 101. *Moody Monthly*, 90(2), 20.
    Discusses the work of Professor Ray Gates in creation care and education. Professor Gates is a faculty member at Grand Rapids Baptist College.
(Topics):popular.

Morud, J. (1989b). Creation groans: are Christians listening? *Moody Monthly*, 90(2), 12-21.
(Topics):popular.

Morud, J. (1989c). Get close to nature. *Moody Monthly*, 90(2), 19.
(Topics):popular, lifestyle.

Morud, J. (1989d). Incomparable masterpiece. *Moody Monthly*, 90(2), 15.
    Focus is on Dr. Calvin DeWitt's work on caring for creation. Dr. DeWitt is Professor of Environmental Studies at the University of Wisconsin (Madison) and Director of Au Sable Institute of Environmental Studies.
(Topics):popular.

Morud, J. (1989e). Keep the warranty. *Moody Monthly*, 90(2), 21.
(Topics):popular, recycling, solid waste.

Moss, R. (1982). *The Earth in our hands*. London: Inter-Varsity Press.
(Topics):popular.

Moss, R. (?, November). A Christian environmental ethic. *Third Way* (London), p. 20-24.
(Topics):popular.

Moss, R. P. (1975). Responsibility in the use of nature (I). *Christian Graduate*, 28(3), 69-80.
(Topics):popular, resources, pollution.

Moss, R. P. (1976). Responsibility in the use of nature (2). *Christian Graduate*, 29(1), 5-14.
(Topics):popular.

Moss, R. P. (1978). Environmental problems and the Christian ethic. In C. F. H. Henry (Ed.), *Horizons of science: Christian scholars speak out* (pp. 66-86). New York: Harper & Row.
(Topics):popular.

Moule, C. F. D. (1967). *Man and nature in the New Testament: some reflections on biblical ecology.* Philadelphia: Fortress Press, Facet Books.
First published in 1964 by The Athlone Press, London.
(Topics):theology.

Mountcastle, W. W., Jr. (1973). The ecological problem from the perspective of comparative religion. *Religious Humanism*, 7(3), 118-125.
A survey of several religious approaches to the creation.
(Topics):popular.

Mpanya, M. (1991). Environmental impacts of a church project in a Zairian village. In G. T. Prance & C. B. DeWitt (Eds.), *Missionary earthkeeping.* Macon, Georgia: Mercer University Press.
(Topics):popular.

Muchina, B. (1989, January - March). God's mirror. *Together, A Publication of World Vision*, p. 12.
(Topics):popular, land, agriculture.

Mundahl, T. W. (1986). From dust to dust: an exploration of elemental integrity. *Word & World*, 6(1), 86-96.
(Topics):land.

Munro, D. W. (1969). Indifference to exploitation unjustifiable. *Journal of the American Scientific Affiliation*, 21(2), 46-47.
(Topics):popular.

Muratore, S. (1983). Journal notes. *Epiphany*, 3(3), 36-39.
(Topics):popular.

Muratore, S. (1985a). Holy weakness, strength of God: from despair to Christian ecology. *Epiphany*, 6(1), 74-77.
(Topics):popular.

Muratore, S. (1985b). Reconciliation with the environment, an estranged realm of the spirit. *Epiphany*, 6(1), 84-88.
(Topics):popular.

Muratore, S. (1985c). Stewardship is enough: ecology as inner priesthood. *Epiphany*, 6(1), 26-34.
(Topics):popular.

Muratore, S. (1985d). Where are the Christians? A call to the church. *Epiphany*, 6(1), 7.
(Topics):popular.

Muratore, S. (1986a). All creation rejoices. *Epiphany*, 7(1), 26-35.
(Topics):popular.

Muratore, S. (1986b). The Earth's end: eschatology and the perception of nature. *Epiphany*, 6(4), 40-49.
(Topics):popular.

Muratore, S. (1987). Ecological ecumenism: inter-faith symbiosis for a healthy planet. In F. W. Krueger (Ed.), *Christian ecology: building an environmental ethic for the twenty-first century* (p. 32). North Webster, Indiana: The North American Conference on Christianity and Ecology.
(Topics):popular.

Muratore, S. (1988). Editorial: the new "Teilhard" at the NACCE: Thomas Berry, the "New Story" at the battle for the Christian mind. *Epiphany*, 8(2), 6-14.
(Topics):popular, creation spirituality.

Murphy, G. (1986). Our fragile ecology. *Catholic Worker*, 53(June-July), 5.
(Topics):popular.

Murphy, G. L. (1982). "Have dominion": the Christian and natural resources. *Journal of the American Scientific Affiliation*, 34(3), 169.
(Topics):resources.

Murphy, N. C. (1985). *Teaching and preaching stewardship: an anthology.* New York: Commission on stewardship, National Council of Churches of Christ in the U. S. A.
(Topics):stewardship.

Murray, A. V. (1956). *Natural religion and Christian theology: an introduction.* New York: Harper.
(Topics):popular book.

Murray, W. (1970). Science and nature and the crisis of contemporary theology. *American Catholic Philosophical Association Proceedings*, 44, 114-121.
(Topics):theology.

Naidoff, B. D. (1978). A man to work the soil: a new interpretation of Genesis 2-3. *Journal for the Study of the Old Testament*, 5, 2-14.
(Topics):theology, land.

Nash, J. A. (1990). Ecological integrity and Christian political responsibility. *Theology & Public Policy*, 1(Fall), 32-48.
(Topics), popular.

Nash, N. (1987). The Buddhist perception of nature project. In S. Davies (Ed.), *Tree of life: Buddhism and protection of nature with a declaration on environmental ethics from his holiness the Dalai Lama and an introduction by Sir Peter Scott* (pp. 31-33). Buddhist Perception of Nature. (Topics):Buddhism.

Nash, R. F. (1989), *The rights of nature: a history of environmental ethics.* Madison, Wis.: The University of Wisconsin Press. 290 pp.
See especially Chapter 4, "The greening of religion."
(Topics):ethics, history.

Nasr, S. H. (1968). *The encounter of man in nature: the spiritual crisis of modern man.* London: George Allen and Unwin, Ltd.
Examines Christianity and nature from the perspective of an Eastern religious background.
(Topics):history.

Naveh, Z. (1989). Neot Kedumim. *Restoration & Management Notes, 7*(1), 9-13.
"Using the Bible as a guide, restorationists in Israel are creating a landscape that evokes the landscape of the prophets."
(Topics):restoration ecology, Israel, Holy Land, Judaism.

Navone, J. (1975). Christian responsibility for the environment. *American Ecclesiastical Review, 169*, 681-689.
(Topics):popular.

Neff, D. (1989). Toro, toro, toro. *Christianity Today, 33*(16), 15.
Short article for the popular press to encourage recycling and composting. The lack of involvement of evangelicals in environmental matters is noted.
(Topics):recycling, solid waste, composting, lifestyle.

Nelson, D. U. (1973). *The theology of ecology: a bibliographical analysis.* M. Div., Western Evangelical Seminary.

(Topics):theology.

Nelson, J. A. (1979). Caring for God's land. *The Lutheran Standard*, 19(June), 32-33.
(Topics):land.

Nelson, J. A. (1984). *Hunger for justice: the politics of food and faith*. Maryknoll, N.Y.: Orbis Books.
(Topics):population, justice, land.

Nelson, J. L., Jr. (1969). *The groanings of creation: an exegetical study of Romans 8:18-27*. Ph.D., Union Theological Seminary.
(Topics):theology.

Nelson, J. R. (1980). *Science and our troubled conscience*. Philadelphia: Fortress Press.
(Topics):book.

Nibley, H. W. (1978). On subduing the Earth. In H. W. Nibley (Ed.), *Nibley on the timely and the timeless*. Salt Lake City, Utah: Publishers Press.
Based on the Mormon faith.
(Topics):book.

Nichols, C. (1987). Faith, politics and ecology: part two: the Environmental Action Foundation. In F. W. Krueger (Ed.), *Christian ecology: building an environmental ethic for the twenty-first century* (p. 84). North Webster, Indiana: The North American Conference on Christianity and Ecology.
(Topics):popular.

Nichols, M. (1987). Eco-feminism and the church: part one: introduction. In F. W. Krueger (Ed.), *Christian ecology: building an environmental ethic for the twenty-first century* (p. 99). North Webster, Indiana: The North American Conference on Christianity and Ecology.
(Topics):ecofeminism.

Nicholson, M. (1970). *The Environmental Revolution.* London: Hodder and Stoughton.
    Considers the Judeo-Christian tradition to have a strong aggressive, exploitative bias toward the environment (see pp. 264f).
    (Topics):book.

Niles, D. P. (1987). Covenanting for justice, peace and the integrity of creation -- an ecumenical survey. *The Ecumenical Review,* 39, 470-484.
    (Topics):ecojustice.

Niles, D. P. (1989). *Resisting the threats to life: covenanting for justice, peace and the integrity of creation.* Geneva: WCC Publications.
    (Topics):popular book.

Nisly, L. L. (1991). People, people everywhere. *ESA Advocate,* 13(5):1-3, 8.
    (Topics):population.

Norman, E. (1979). *Christianity and the world order.* New York: Oxford University Press.
    (Topics):popular book.

North Carolina (Land Stewardship Council) (1986). *Stewardship of the Land in North Carolina.* Pittsboro, North Carolina: The Land Stewardship Council of North Carolina.
    Available through Land Stewardship Council, Route 4, Box 426, Pittsboro, North Carolina 27312.
    (Topics):land.

North, R. (1954). *Sociology of the biblical jubilee.* Rome: Pontifical Biblical Institute.
    (Topics):theology.

Nurnberger, K. (Ed.) (1987). *Ecology and Christian ethics in a semi-industrialised and polarised society.* Pretoria, South Africa: University of South Africa.

(Topics):ethics.

Oates, D. (1983). Whose is the Earth? *Other Side*, 143(August), 14-17.
(Topics):popular.

Obayashi, H. (1973). Nature and historicization: a theological reflection on ecology. *Cross Currents*, 23, 140-152.
(Topics):theology, history.

Oegema, T. (1987). Pressures in the agri-food system in the year 2000. *Earthkeeping: A Quarterly on Faith and Agriculture*, 3(1), 4-7.
(Topics):land, agriculture.

Ogg, W., & Butcher, W. (1962). Economic resources and population. *Journal of the American Scientific Affiliation*, 14(1), 15-16.
(Topics):population.

Ogle, G. (1988, November). McFarland's children: "Who is next?": "There is no scientific evidence to indict pesticides." *Christian Social Action*, pp. 22-23.
(Topics):pesticides, pollution.

Oliver, D. F. (1987). 'God's rest' the core and Leitmotif of a Christian holistic view of reality? In W. S. Vorster (Ed.), *Are We Killing God's Earth?* (pp. 100-118). Pretoria, South Africa: University of South Africa.
(Topics):theology.

Ollenburger, B. C. (1987). Isaiah's creation theology. *Ex Auditu*, 3, 54-71.
(Topics):theology.

Olsen-Tjensvold, I. (1978). *Response to creation: Christian environmentalism and the theology and ethics of Richard Niebuhr*. Ph.D., Syracuse University.
(Topics):theology.

Olsen-Tjensvold, I. (1986). A trinity we can live with. *The Egg: A Journal of Eco-Justice*, 6(1), 9-10.
(Topics):popular, ethics.

Olson, D. T. (1986). Biblical perspectives on the land. *Word & World*, 6(1), 18-27.
(Topics):land, theology.

Olson, J. (1987). Ecology: legal rights and Christian concerns. In F. W. Krueger (Ed.), *Christian ecology: building an environmental ethic for the twenty-first century* (pp. 78-80). North Webster, Indiana: The North American Conference on Christianity and Ecology.
(Topics): popular, law, judicial system.

Orgon, J. (1987). Christian ecology in Europe: part two: the continental perspective. In F. W. Krueger (Ed.), *Christian ecology: building an environmental ethic for the twenty-first century* (p. 68). North Webster, Indiana: The North American Conference on Christianity and Ecology.
(Topics):popular.

Orgon, J., & Artaza, M. P. (1987). Faith, politics, and ecology: part one: justice for creation. In F. W. Krueger (Ed.), *Christian ecology: building an environmental ethic for the twenty-first century* (p. 83). North Webster, Indiana: The North American Conference on Christianity and Ecology.
(Topics):popular.

Osei-Mensah, G. (1980). The Christian life-style. In J. Stott & R. Coote (Eds.), *Down to Earth: studies in Christianity and culture*. Grand Rapids, MI: Eerdmans Publishing Company.
(Topics):lifestyle.

Ostendorf, D. L. (1986). Toward wholeness and community: strategies for pastoral and political response to the American rural crisis. *Word & World*, 6(1), 55-65.
(Topics):land.

Oswalt, J. N. (1971). Is Christianity Responsible for This Mess? *Eternity*, 22(9), 17-18.
(Topics):popular.

Overby, R. (1987). Christian ecological economics: part one: "planetheonomics." In F. W. Krueger (Ed.), *Christian ecology: building an environmental ethic for the twenty-first century* (pp. 70-71). North Webster, Indiana: The North American Conference on Christianity and Ecology.
(Topics):economics.

Overman, R. H. (1971). A Christological view of nature. *Religious Education*, 66(1), 36-44.
(Topics):popular.

Owens, O. D. (1974). Salvation and ecology: missionary imperatives in light of a new cosmology. *Foundations*, 17, 106-123.
(Topics):popular.

Owens, O. D. (1976). Interdependence. *Foundations*, 19, 196-202.
(Topics):popular.

Owens, O. D. (1977). *Stones into bread? What does the Bible say about feeding the hungry today.* Valley Forge: Judson Press.
(Topics):agriculture.

Owens, V. S. (1976). Go to the garden where decay redeems. *Christianity Today*, 21(6), 342-344.
(Topics):popular.

Owens, V. S. (1978). Consider the fingerprints of God: before nature one must be silent and stare. *Christianity Today*, 23(4), 226-229.
(Topics):popular.

Owens, V. S. (1983a). *And the trees clap their hands: faith, perception, and the new physics.* Grand Rapids, MI: Eerdmans.
    The focus is on physics, but a clear statement of man's relation to and responsibility for nature is made.
    (Topics):popular book.

Owens, V. S. (1983b). Where in the world? *The Reformed Journal*, 33(9), 8-9.
    (Topics):popular.

Owens, W. D. (1989). The eco-justice working group. *The Egg: A Journal of Eco-Justice*, 9(3), 3, 16.
    (Topics):popular, ecojustice.

Paddock, J., Paddock, N., and Bly, C. (1986). *Soil and survival: land stewardship and the future of American agriculture.* San Francisco: Sierra Club Books. 217 pp.
    See especially chapter 8: "Dust to dust: land in the Jewish and Christian traditions."
    (Topics):land.

Paetkau, P., Harder, G., and Sawatzky, D. (1978). *God, man, land, interrelationship programs for camps.* Newton, Kan.: Faith and Life Press.
    (Topics):education.

Palmer, M., Nash, A., & Hottingh, I. (1987). *Faith and Nature.* London: Rider Books.
    (Topics):book.

Pang, C. C. (1983). The significance of Luther's thought on nature in the Christian witness of Asia today. *East Asia Journal of Theology*, 1(1), 1-14.
    (Topics):theology.

Paradise, S. (1979). Visions of the good society and the energy debate. *Anglican Theological Review*, 61(1), 106-117.
    (Topics):energy.

Park, B. (1987). Tilling and keeping. *Earthkeeping: A Quarterly on Faith and Agriculture*, 3(3), 17-18.
(Topics):land.

Parmenov, A. J. (1989). Christian responsibility for the environment. *The Journal of the Moscow Patriarchate*, 3, 38-39.
From the Church and Society meeting, Tambov, Russia, Sept 18-23, 1988.
(Topics):popular.

Passmore, J. (1974). *Man's responsibility for nature*. New York: Charles Scribner's Sons.
(Topics):book, ethics.

Passmore, J. (1975). Attitudes to nature. In R. S. Peters (Ed.), *Nature and Conduct, Royal Institute of Philosophy Lectures* (pp. 251-264). London: Macmillan.
(Topics):history.

Passmore, J. (1977). Ecological problems and persuasion. In G. Dorsey (Ed.), *Equality and Freedom*. Dobbs Ferry, NY: Oceana Publications.
(Topics):popular.

Pater, C. (1989). The battle of the bulk. *Earthkeeping: A Quarterly on Faith and Agriculture*, 5(4/5), 25.
In the journal section under "practicing stewardship." Gives some practical applications.
(Topics):solid waste.

Paternoster, M. (1976). *Man: the world's high priest: an ecological approach*. Oxford: SLG (Sisters of the Love of God) Press.
(Topics):book.

Paternoster, M. C. (1971). Thy Humbler Creation. *Church Quarterly*, 3(January), 204-214.
(Topics):popular.

Patey, E. H. (1976). *Christian lifestyle*. London: Mowbrays.
(Topics):lifestyle.

Pattison, E. M. (1974). Will means destroy ends? *Journal of the American Scientific Affiliation*, 26(1), 20.
(Topics):population.

Peacocke, A. R. (1961). The Christian faith in a scientific age. *Religion in Education*, 28, 53-59.
(Topics):stewardship, popular.

Peacocke, A. R. (1979). *Creation and the world of science*. Oxford: Oxford University Press.
(Topics):theology.

Peacocke, A. R. (1987). Rethinking religious faith in a world of science. In F. T. Birtel (Ed.), *Religion, science and public policy* (pp. 3-29). New York: The Crossroad Publishing Company.
(Topics):history, theology.

Pearson, F. (1974). How the energy crisis can keep us in moral fighting trim. *The Christian Century*, 91(9), 256-259.
(Topics):energy.

Pelcovitz, R. (1970). Ecology and Jewish theology. *Jewish Life*, 37, 23-32.
(Topics):Judaism, theology.

Peters, K. E. (1977). Realities and ideals in the world system. *Zygon*, 12(2), 169-174.
(Topics):popular.

Peters, K. E. (1987). Toward a physics, metaphysics, and theology of creation: a trinitarian view. In F. T. Birtel (Ed.), *Religion, science, and public policy* (pp. 96-112). New York: The Crossroad Publishing Company.
(Topics):theology.

Peters, T. (1987). A hole in the heavens. *Dialog*, 26, 8.

(Topics):pollution, ozone.

Pfrimmer, D. (1988). The future of farming and farming the future. *Consensus,* 14(2), 49-59.
(Topics):land, agriculture.

Philibert, P. J. (1982). Becoming human in the world for the sake of the world. *Epiphany,* 2(3), 20-27.
(Topics):popular.

Phillips, R. (1973). The Tree Managers. *Church and Society,* 64(2), 44-61.
(Topics):resources.

Phipps, W. E. (1988). Asa Gray's theology of nature. *American Presbyterians: Journal of Presbyterian History,* 66, 167-175.
(Topics):theology.

Pignone, M. M. (1978). Concentrated ownership of land. In M. E. Jegen & B. V. Manno (Eds.), *The Earth is the Lord's: essays on stewardship* (pp. 112-129). New York: Paulist Press.
(Topics):popular, land.

Pippert, W. G. (1971). What's your EQ? *Christian Life,* 33(July), 20+.
(Topics):popular.

Pitchford, J. (1989, January - March). In tune with God and his environment. *Together, A Publication of World Vision,* p. 11.
(Topics):popular.

Plantinga, C., Jr. (1989). Signals from Seoul. *The Reformed Journal,* 39(10), 10-13.
Report on the 22nd General Council of the World Alliance of Reformed Churches (WARC) at Yonsei University, Seoul, Korea on August 15-27, 1989.
(Topics):popular.

Platman, R. H. (1971). Theology and ecology: a problem for religious education. *Religious Education*, 66(1), 14-23.
(Topics):theology.

Platt, D. R. (1982). Stewardship of natural ecosystems: a case study in the tallgrass prairie. In E. R. Squiers (Ed.), *The environmental crisis: the ethical dilemma* (pp. 115-134). Mancelona, MI: Au Sable Institute of Environmental Studies.
(Topics):resources, land.

Pobee, J. S. (1977). Towards a theology and ethics of population dynamics. In J. S. Bebee (Ed.), *Religion, morality, population dynamics*. Chapel Hill: Popular Center.
(Topics):population.

Pobee, J. S. (1985). Creation faith and responsibility for the world. *Journal of Theology for Southern Africa*, 50(March), 16-26.
(Topics):theology.

Pobee, J. S. (1990). Lord, creator-spirit, renew and sustain the whole creation: some missiological perspectives. *International Review of Mission*, 79(314), 151-158.
    Part of a World Council of Churches publication on environmental issues.
(Topics):popular.

Points, G. P. (1975). Nature-man-God: a post-modern view. *Lexington Theological Quarterly*, 10, 11-22.
(Topics):theology.

Pollard, N. (1984). The Israelites and their environment. *The Ecologist*, 14(3), 125-133.
(Topics):popular.

Pollard, W. G. (1969). Man on a spaceship. *Journal of the American Scientific Affiliation*, 21(2), 34-39.
(Topics):popular.

Pollard, W. G. (1970). God and his creation. In M. Hamilton (Ed.), *This little planet* (pp. 43-76). New York: Charles Scribner's Sons.
(Topics):popular.

Pollard, W. G. (1980a). Not an avoidable problem. *Journal of the American Scientific Affiliation*, 32(2), 88.
(Topics):nuclear.

Pollard, W. G. (1980b). A theological view of nuclear energy. *Journal of the American Scientific Affiliation*, 32(2), 70-74.
(Topics):nuclear.

Polman, G. (1987a). CFFA convention report: an encounter with change. *Earthkeeping: A Quarterly on Faith and Agriculture*, 3(1), 16-17.
Discussion of biotechnology.
(Topics):genetic engineering.

Polman, G. (1987b). Close Encounters of a Powerful Kind. *Earthkeeping: A Quarterly on Faith and Agriculture*, 3(2), 5-9.
Report of a group of Christian farmers on a study tour to Mexico.
(Topics):land.

Polman, G. (1987c). Developing links to share in creation. *Earthkeeping: A Quarterly on Faith and Agriculture*, 3(2), 3.
(Topics):land.

Polman, G. (1988). Of mice and men and rights. *Earthkeeping: A Quarterly on Faith and Agriculture*, 4(1), 3.
(Topics):animal rights.

Polman, G. (1989). Have a heart. *Earthkeeping: A Quarterly on Faith and Agriculture*, 5(2), 3.

Editorial urging readers to be earthkeepers.
(Topics):popular.

Pope, R. M. (1972). South Elkhorn: the church and a theology of nature. *Lexington Theological Quarterly*, 7, 113-120.
(Topics):theology.

Porter, H. B. (1986). *A song of creation*. Cambridge, MA: Cowley Publications.
(Topics):popular book.

Potter, P. (1985). Development as an ecumenical challenge. *Earthkeeping: A Quarterly on Faith and Agriculture*, 1(2), 16-18.
(Topics):popular.

Prance, G. T. (1990). Giver of life -- sustain your creation. *International Review of Mission*, 79(April), 159-166.
(Topics):popular, biodiversity.

Prance, G. T. (1991). The ecological awareness of the Amazon Indians. In G. T. Prance and C. B. DeWitt (Eds.), *Missionary earthkeeping*. Macon, Georgia: Mercer University Press.
(Topics):popular book, ethnobotany.

Prance, G. T., & DeWitt, C. B. (Eds.) (1991). *Missionary Eathkeeping*. Macon, Georgia: Mercer University Press.
(Topics):popular book.

Presbyterian Church (U.S.A.) (1989). *Keeping and healing the creation*. Louisville: Eco-Justice Task Force, Committee on Social Witness Policy, Presbyterian Church (U.S.A.).
Principal, but unlisted, author is W. E. Gibson.
(Topics):book, ecojustice.

Preston, R. H. (1980). The question of a just, participatory and sustainable society. *Bulletin of the John Rylands University Library of Manchester*, 63(1), 95-117.
(Topics):popular.

Primavesi, A. (1990). The part for the whole? An ecofeminist enquiry. *Theology*, 93(Sept./Oct.), 355-362.
(Topics):theology, ecofeminism.

Przewozny, B. (1988). Integrity of creation: a missionary imperative. *SEDOS Bulletin*, 11(Dec. 15), 363-374.
(Topics):popular.

Rae, E. (1987). The feminine image of God and its implications for women, men and the Earth. In F. W. Krueger (Ed.), *Christian ecology: building an environmental ethic for the twenty-first century* (p. 98). North Webster, Indiana: The North American Conference on Christianity and Ecology.
(Topics):feminism, popular.

Raffensperger, C. (1990). All God's critters got a place in the choir. *Daughters of Sarah*, 16(3), 4-6.
(Topics):ecofeminism.

Rajotte, F. (1988, Sept.-Oct.). Creation theology at the W.C.C. *Ecumenist*, p. 85-90.
(Topics):popular.

Rajotte, F. (1990). Justice, peace and the integrity of creation. *Religious Education*, 85(1), 5-14.
(Topics):ecojustice, popular

Rakestraw, R. (1986). The contribution of John Wesley toward an ethics of nature. *Drew Gateway*, 56(3), 14-25.
(Topics):ethics.

Ramm, B. (1971). Evangelical theology and technological shock. *Journal of the American Scientific Affiliation*, 23(2), 52-56.
(Topics):genetic engineering.

Ramon, J., & Bube, R. H. (1985). Appropriate technology for the third world. *Journal of the American Scientific Affiliation*, 37(2), 66-71.
(Topics):technology, appropriate technology.

Ranck, L. (Ed.) (1989). Environmental justice issues. *Christian Social Action*, 2, 4-24.
(Topics):ecojustice.

Ranck, L. (Ed.) (1990). [Methodist response to environmental crisis]. *Christian Social Action*, 3(September), 4-16,25-32,35-36,39.
(Topics):popular.

Rands, B. (1989, January - March). Beyond Timbuktu, oases of hope. *Together, A Publication of World Vision*, pp. 4-5.
(Topics):popular.

Rasmussen, L. (1975). The future isn't what it used to be: "limits to growth" and Christian ethics. *Lutheran Quarterly*, 27(2), 101-111.
(Topics):ethics.

Rasmussen, L. (1985). The bishops and the economy: three appraisals; III. On Creation, on growth. *Christianity and Crisis*, 45(19), 473-476.
(Topics):economics.

Rasmussen, L. (1987). Creation, church and Christian responsibility. In W. Granberg-Michaelson (Ed.), *Tending the garden: Essays on the Gospel and the Earth* (pp. 114-131). Grand Rapids, MI: Eerdmans Publishing Company.
(Topics):theology, popular.

Rasmussen, L. L. (1990). The planetary environment: challenge on every front. *Theology & Public Policy*, 2(Summer), 3-14.
(Topics):popular.

Raven, C. E. (1940). *The gospel and the church: a study of distortion and its remedy.* New York: Charles Scribner's Sons.
(Topics):history.

Raven, C. E. (1953). *Science and religion.* Cambridge: Cambridge University.
(Topics):history, theology.

Raven, C. E. (1955). *Christianity and science.* London: Lutterworth Press.
(Topics):theology.

Ravid, F. (1987). Kebash: the marital commandment to subdue the Earth. *Epiphany,* 7, 66-70.
(Topics):popular.

Redekop, C. (1986). Toward a Mennonite theology and ethic of creation. *Mennonite Quarterly Review,* 60(3), 387-403.
(Topics):theology.

Regan, R. (1989). A river too good to waste. *Christian Social Action,* 2, 14-15.
(Topics):resources, conservation.

Regenstein, L. G. (1991), *Replenish the Earth: a history of organized religion's treatment of animals and nature -- including the Bible's message of conservation and kindness toward animals.* New York: Crossroad.
(Topics):history, animal rights, popular book, Judaism, Hinduism, Jainism, Buddhism, Islam, the Baha'i faith.

Rehwaldt, A. C. (1969). The Christian world view and the new era in science. *Concordia Theological Monthly,* 40(1), 13-23.
(Topics):theology.

Reid, L. (1975). Towards a religion of the environment. *The Teilhard Review,* 10, 16-19.

(Topics):popular.

Reidel, C. H. (1971). Christianity and the environmental crisis. *Christianity Today*, 15(15), 684-688.
(Topics):popular.

Reinhart, P. (1983). The Bridge. *Epiphany*, 3(3), 21.
(Topics):popular.

Reinhart, P. (1985a). Editorial introducing *Epiphany* issue 6(1) titled "To be Christian is to be Ecologist." *Epiphany*, 6(1), 1-2.
(Topics):popular.

Reinhart, P. (1985b). Eternal festival: folk culture, celebrations, and Earth stewardship. *Epiphany*, 6(1), 46-50.
(Topics):popular.

Reinhart, P. (1985c). The ten talents of stewardship and the angelic dimension. *Epiphany*, 6(1), 36-43.
(Topics):popular.

Reinhart, S. (1987). The Christian family: foundation for environmental renewal: part two: family and the formation of a Christian ecologist. In F. W. Krueger (Ed.), *Christian ecology: building an environmental ethic for the twenty-first century* (pp. 89-90). North Webster, Indiana: The North American Conference on Christianity and Ecology.
(Topics):popular.

Rendtorff, R. (1979). 'Subdue the Earth': man and nature in the Old Testament. *Theology Digest*, 27(3), 213-216.
(Topics):popular.

Reumann, J. (1973). *Creation and new creation: the past, present, and future of God's creative activity.* Minneapolis: Augsburg Publishing House.
(Topics):theology.

Review and Expositor (thematic issue) (1972). Ecology and the church. *Review and Expositor: A Baptist Theological Journal*, 69(1), 3-76.
(Topics):theology.

Reynolds, S. (1987, April-June). Too many people, not enough land. *Together: A Publication of World Vision*, p. 3.
(Topics):land, population.

Richard, L. (1986). Toward a renewed theology of creation: implications of human rights. *Eglise et Théologie*, 17(2), 149-170.
(Topics):theology.

Richardson, C. C. (1972). A Christian Approach to Ecology. *Religion in Life*, 41(4), 462-479.
(Topics):history, popular.

Rifkin, J. (1987). Rethinking our world view: alternative futures for the 21st century. In F. W. Krueger (Ed.), *Christian ecology: building an environmental ethic for the twenty-first century* (pp. 60-63). North Webster, Indiana: The North American Conference on Christianity and Ecology.
(Topics):world view, popular.

Rifkin, J., & Howard, T. (1979). *The emerging order: God in the age of scarcity*. New York: Putnam's Sons.
(Topics):economics.

Riga, P. J. (1970). Ecology and theology. *The Priest*, 26, 16-21.
(Topics):popular.

Rigdon, V. B. (1983). Creation within the love of God. *Journal of Presbyterian History*, 61(Spring), 43-54.
(Topics):history, theology.

Riggs, D. (1987). Wilderness retreats: part two: wilderness and prayer. In F. W. Krueger (Ed.), *Christian ecology:*

building an environmental ethic for the twenty-first century
(p. 56). North Webster, Indiana: The North American
Conference on Christianity and Ecology.
(Topics):wilderness.

Rimbach, J. A. (1987). All creation groans: theology/ecology
in St. Paul. *Asia Journal of Theology,* 1(2), 379-391.
(Topics):theology.

Riverson, J. J. (1986, July-September). Water and chil-
dren's health: the view from Ghana. *Together: a Journal of
World Vision International,* pp. 10-13.
(Topics):water, health.

Robb, J. W. (1981). The Christian and the new biology.
Encounter, 42, 197-205.
(Topics):genetic engineering.

Robbins, J. K. (1987). The environment and thinking about
God. *Encounter,* 48(4), 401-415.
(Topics):popular.

Robbins, J. W. (1972). Ecology: the abolition of man.
Creation Research Society Quarterly, 8(4), 280-284.
    A highly critical paper attacking Klotz 1971,
"Creationism and our ecological crisis," in which the
author asserts that there is no ecological crisis and
Christians need not be concerned.
(Topics):popular.

Robbins, W. W. (1970). The theological values of life and
nonbeing. *Zygon,* 5(4), 339-352.
(Topics):popular.

Robertson, J. (1989). A new economics for the 21st century.
The Egg: A Journal of Eco-Justice, 9(1), 4-5.
(Topics):economics.

Robinson, H. W. (1946). *"God and nature." Part I in inspiration and revelation in the New Testament.* Oxford: Clarendon Press.
(Topics):theology.

Robinson, W. D. (1973). Theology of only one Earth. *The Expository Times,* 84(September), 355-358.
(Topics):theology.

Rogers, D. B. (1990). Coyotes and sheep: what the exile teaches the Church about ecological education. *Journal of Theology* (United Theological Seminary), 94, 7-19.
(Topics):theology, education.

Rondeel, P. (1989, January - March). Those who destroy the Earth. *Together, a publication of World Vision,* p. 3.
(Topics):popular.

Root, M. J. (1979) *Creation and redemption: a study of their interrelationship, with special reference to the theology of Regin Prenter.* Ph.D., Yale University.
(Topics):theology.

Ross, E. M. (1987). *Humans' creation in God's image: Richard of St. Victor, Walter Hilton, and contemporary theology.* Ph.D., The University of Chicago.
(Topics):theology.

Rossi, A. (1989). Hearing the word of God in creation. *Firmament: The Quarterly of Christian Ecology,* 1(3), 2-3, 16.
Also published in 1989, *Epiphany,* 9(4)74-76.
(Topics):popular.

Rossi, A. (Ed.) (1990). For the transfiguration of nature: papers from the symposium on orthodoxy and ecology. *Epiphany,* 10(Spring), 7-74.
(Topics):popular.

Rossi, V. (1981a). An Open Letter to the Editors of "The Next Whole Earth Catalogue." *Epiphany*, 2(1), 16-31.
(Topics):popular.

Rossi, V. (1981b). The Eleventh Commandment: toward an ethic of ecology. *Epiphany*, 1(4), 3-19.
(Topics):popular.

Rossi, V. (1985a). The Earth is the Lord's: excerpts from "The eleventh commandment: toward an ethic of ecology." *Epiphany*, 6(1), 3-6.
(Topics):popular.

Rossi, V. (1985b). Theocentrism: the cornerstone of Christian ecology. *Epiphany*, 6(1), 8-14.
(Topics):popular.

Rossi, V. (1987). Theocentrism: the cornerstone of Christian ecology. In F. W. Krueger (Ed.), *Christian ecology: building an environmental ethic for the twenty-first century* (pp. 27-29). North Webster, Indiana: The North American Conference on Christianity and Ecology.
(Topics):popular.

Rossi, V. (1988). Christian ecology is cosmic Christology. *Epiphany*, 8(Winter), 52-62.
(Topics):popular.

Rounder, L. S. (Ed.) (1984). *On Nature*. Notre Dame: University of Notre Dame Press.
(Topics):history.

Rowald, H. L. (1977) *The theology of creation in the Yahweh speeches of the book of Job as a solution to the problem posed by the book of Job*. Th.D., Concordia Seminary in Exile, St. Louis.
(Topics):theology.

Royer, W. (1987). Nature mysticism in St. Seraphim of Sarov and St. John of Kronstadt. In F. W. Krueger (Ed.),

Christian ecology: building an environmental ethic for the twenty-first century (pp. 52-53). North Webster, Indiana: The North American Conference on Christianity and Ecology.
(Topics):popular.

Rudin, A. (1986). A very northern church. *Engage / Social Action*, 14(4), 29-31.
(Topics):energy, conservation.

Rudisell, R. T. (1973) *Contemporary Representative American approaches to a theology of environment.* Th.D., Southwestern Baptist Theological Seminary.
(Topics):theology.

Ruether, R. R. (1972). *Liberation theology: human hope confronts Christian history and American power.* New York: Paulist Press.
A general discussion of liberation theology. Contains some material dealing with the environment.
(Topics):theology.

Ruether, R. R. (1974). Rich nations/poor nations and the exploitation of the earth. *Dialog* (St. Paul, Minn.), 13(Summer), 201-207.
(Topics):popular.

Ruether, R. R. (1975). *New woman, new earth: sexist ideologies and human liberation.* New York: The Seabury Press.
Feminist emphasis with some references to ecology and nature.
(Topics):theology, ecofeminism.

Ruether, R. R. (1978). The biblical vision of the ecological crisis. *Christian Century*, 95(38), 1129-1132.
(Topics):popular.

Ruether, R. R. (1981). Ecology and human liberation: a conflict between the theology of history and the theology of na-

ture. In R. R. Ruether (Ed.), *To change the world: Christology and cultural criticism*. New York: Crossroad.
(Topics):theology.

Ruether, R. R. (1983). *Sexism and God-talk: toward a feminist theology*. Boston: Beacon Press.
(Topics):ecofeminism, theology.

Russell, C. A. (1985). *Crosscurrents: interactions between science and faith*. Grand Rapids, MI: Eerdmans Publishing Company.
See especially chapter 11, "Polluting effluent: science and the environment."
(Topics):pollution, popular.

Rust, E. C. (1953). *Nature and man in Biblical thought*. London: Lutterworth Press.
(Topics):theology.

Rust, E. C. (1967). *Science and faith: towards a theological understanding of nature*. New York: Oxford University Press.
(Topics):theology.

Rust, E. C. (1971). *Nature -- garden or desert? An essay in environmental theology*. Waco, TX: Word Books.
(Topics):theology.

Rust, E. C. (1972). Nature and man in theological perspective. *Review and Expositor: A Baptist Theological Journal*, 69(1), 11-22.
(Topics):theology.

Rust, E. C. (1980). A Christian understanding of and attitude toward nature. In P. Simmons (Ed.), *Issues in Christian ethics*. Nashville, Tenn.: Broadman Press.
(Topics):ethics, theology.

Ryan, P. (1987). A Christian eco-channel design for local television. In F. W. Krueger (Ed.), *Christian ecology:*

building an environmental ethic for the twenty-first century (p. 84). North Webster, Indiana: The North American Conference on Christianity and Ecology.
(Topics):popular.

Ryan, T. (Ed.) (1990). Come Holy Spirit, renew the whole creation. *Ecumenism*, 100(December), 3-27.
(Topics):popular.

Rybeck, W., & Pasquariello, R. D. (1987). Combating modern-day feudalism: land as God's gift. *The Christian Century*, 104(16), 470-472.
(Topics):land, economics.

Rzadki, J. A., & Hilts, S. G. (1989). Partners in conservation: voluntary co-operation is replacing regulation among government ministries and private landowners. *Earthkeeping: A Quarterly on Faith and Agriculture*, 5(4/5), 22-23.
(Topics):conservation, land, agriculture, popular.

Sagan, C. (1990). Guest comment: preserving and cherishing the Earth -- an appeal for joint commitment in science and religion. *American Journal of Physics*, 58(7), 615.
A letter by Sagan and signed by ten leading U.S. and foreign scientists calling for the religious community to join in the effort to preserve creation. The appeal drew both positive and negative comment from the scientific community and several letters were published in the *American Journal of Physics*, 58(12):1127-1128. One letter was by Howard Van Till of Calvin College.
(Topics):religious commitment.

Samsonov, I. J. (1988). Modern ecological crisis in the light of the Bible and Christian world view. *The Journal of the Moscow Patriarchate*, 11, 40-45.
(Topics):popular.

Santmire, H. P. (1966) *Creation and nature: a study of the doctrine of nature with special attention to Karl Barth's doctrine of creation.* Th.D., Harvard University.
(Topics):theology.

Santmire, H. P. (1968a). I-thou, I-it, and I-ens. *Journal of Religion*, 47, 260-273.
(Topics):theology.

Santmire, H. P. (1968b). New theology of nature. *Lutheran Quarterly*, 20, 290-308.
(Topics):theology.

Santmire, H. P. (1969). The integrity of nature. In A. Stefferud (Ed.), *Christians and the good Earth.* New York: Friendship Press.
(Topics):popular.

Santmire, H. P. (1970a). *Brother Earth: nature, God, and ecology in a time of crisis.* New York: Thomas Nelson, Inc.
(Topics):history, popular.

Santmire, H. P. (1970b). Ecology and schizophrenia: historical dimensions of the American crisis. *Dialog*, 9(Summer), 175-192.
(Topics):history.

Santmire, H. P. (1970c). The struggle for an ecological theology: a case in point. *Christian Century*, 87, 275-277.
(Topics):popular.

Santmire, H. P. (1973). Reflections on the alleged ecological bankruptcy of western theology. In E. Steffenson,W. J. Herrscher, & R. S. Cook (Eds.), *Ethics for environment: three religious strategies.* Green Bay, WI: UWGB Ecumenical Center.
(Topics):theology.

Santmire, H. P. (1975). Reflections on the alleged ecological bankruptcy of western theology. *Anglican Theological Review*, 57(April), 131-152.
> This paper is a modified form of Santmire 1973.
> (Topics):theology.

Santmire, H. P. (1976). Ecology, justice, and theology: beyond the preliminary skirmishes. *The Christian Century*, 93(17), 460-464.
> (Topics):popular.

Santmire, H. P. (1977). Ecology and ethical ecumenics. *Anglican Theological Review*, 59(1), 98-102.
> (Topics):popular, theology.

Santmire, H. P. (1980). St. Augustine's theology of the biophysical world. *Dialog*, 19(3), 174-185.
> Reprinted in 1983, *Epiphany*, 3(3):52-61.
> (Topics):history, theology.

Santmire, H. P. (1982). Studying the doctrine of creation: the challenge. *Dialog*, 21(Summer), 195-200.
> (Topics):theology.

Santmire, H. P. (1984). The future of the cosmos and the renewal of the church's life with nature. *Word and World*, 4(4), 410-421.
> (Topics):theology.

Santmire, H. P. (1985a). The liberation of nature: Lynn White's challenge anew. *The Christian Century*, 102(18), 530-533.
> (Topics):popular.

Santmire, H. P. (1985b). *The travail of nature: the ambiguous ecological promise of Christian theology*. Philadelphia: Fortress Press.
> (Topics):theology.

Santmire, H. P. (1986). Toward a new theology of nature. *Dialog*, 25(1), 43-50.
(Topics):theology.

Sauer, E. (1962). *The king of the Earth.* Grand Rapids, MI: Wm. B. Eerdmans.
(Topics):theology.

Schaeffer, F. (1970). *Pollution and the death of man.* Wheaton, Ill.: Tyndale House.
(Topics):book.

Schaeffer, F. (1983). Christian stewardship. *Epiphany*, 3(3), 22-25.
(Topics):popular.

Schaeffer, J. (1987). Energy conservation and the local church: the energy task force of the Archdiocese of Milwaukee. In F. W. Krueger (Ed.), *Christian ecology: building an environmental ethic for the twenty-first century* (p. 102). North Webster, Indiana: The North American Conference on Christianity and Ecology.
(Topics):energy, conservation.

Schaffer, A. (1982). The agricultural and ecological symbolism of the four species of sukkot. *Tradition*, 20, 128-140.
(Topics):Judaism, agriculture.

Schecter, J. (1981) *The theology of the land in Deuteronomy.* Ph.D., University of Pittsburgh.
(Topics):theology, land.

Scheffczyk, L. (1970). *Creation and providence.* New York: Herder.
(Topics):theology.

Scherer, D. (Ed.) (1973). *Earth ethics for today and tomorrow: responsible environmental trade-offs.* Bowling Green, Ohio: Bowling Green State University Environmental Studies Center.

Conference Proceedings.
(Topics):ethics.

Schicker, G. E. (1988a). An institute for 'earth keeping.' *The Christian Century*, 1-5(26), 808-812.
(Topics):popular.

Schicker, G. E. (1988b). Tending the garden as God's stewards. *Eternity*, 39(9), 68-69.
(Topics):popular.

Schild, M. (1983). Nature and the natural in Luther's thought: reaction to a paper by Yoshikazu Tokusen. *East Asia Journal of Theology*, 1(1), 38-45.
(Topics):theology.

Schillebeeckx, E. (1980). 'All is grace.' Creation and grace in the Old and New Testaments. In E. Schillebeeckx (Ed.), *Christ: the experience of Jesus as Lord* (pp. 515-530). New York: Seabury Press.
(Topics):theology.

Schillebeeckx, E. (1981). Kingdom of God: creation and salvation. In E. Schillebeeckx (Ed.), *Interim Report on the Books Jesus and Christ* (pp. 105-124). New York: Seabury Press.
See especially chapter six, "Kingdom of God: Creation and salvation."
(Topics):theology.

Schilling, H. K. (1973). *The new consciousness in science and religion*. Philadelphia: United Church.
(Topics):theology.

Schlegel, C. (1989). Making ecology relevant to Christianity--on the farm. *Earthkeeping: A Quarterly on Faith and Agriculture*, 5(3), 11.
(Topics):agriculture, land.

Schmemann, A. (1963). *For the life of the world.* New York: National Student Christian Federation.
　　Eastern Orthodox tradition.
　　(Topics):theology.

Schmitz, B. G. (1990). The eucharist and ecology: a sacramental look at creation. *Daughters of Sarah,* 16(3), 11.
　　(Topics):ecofeminism.

Schumacher, E. F. (1973). *Small is beautiful.* New York: Harper & Row.
　　Is now considered to be a classic in the growing controversy over the role and future of high technology in our society. He questions the assumption that bigger is better and convincingly discusses the role of intermediate technology.
　　(Topics):lifestyle, technology, appropriate technology, popular book.

Schumacher, E. F. (1974). Message from the universe. *Resurgence,* 5(5), 6-8.
　　(Topics):popular.

Schumacher, E. F. (1977). *A guide for the perplexed.* New York: Harper & Row.
　　(Topics):lifestyle.

Schumacher, E. F. (1979). *Good work.* New York: Harper & Row.
　　(Topics):lifestyle.

Schumacher, E. F. (1985). Excerpts from "Message from the universe." *Epiphany,* 6(1), 56-58.
　　(Topics):popular.

Schumaker, M. (1973). Nature's servant-king. In *Genesis & ecology: an exchange* (pp. 13-26). Kingston, Ontario: Queen's Theological College.
　　See comments on Gowan and Schumaker 1973.
　　(Topics):popular.

Schutz, L. H. (1991, May 20). Environmentalism for good and ill. *Focus on the Family Citizen*, 12-13.
    (Topics):population.

Schuurman, D. J. (1987). Creation, eschaton, and ethics: an analysis of theology and ethics in Jurgen Moltmann. *Calvin Theological Journal*, 22, 42-67.
    (Topics):theology, ethics.

Schwartz, R. H. (1984). *Judaism & global survival*. New York: Vintage Press.
    (Topics):Judaism, book.

Schwarz, H. (1974). Eschatological dimensions of ecology. *Zygon*, 9(December), 323-338.
    (Topics):popular.

Schwarz, H. (1977). *Our cosmic journey: Christian anthropology in the light of current trends in the sciences, philosophy and theology*. Minneapolis, Minn.: Augsburg Publishing House.
    (Topics):theology, books.

Schwarz, H. (1979a). Good and evil in technology as a question of Christian values. *Journal of the American Scientific Affiliation*, 31(4), 205-209.
    (Topics):technology.

Schwarz, H. (1979b). *On the way to the future: a Christian view of eschatology in the light of current trends in religion, philosophy, and science*, rev. ed. Minneapolis: Augsburg Publishing House.
    (Topics):theology.

Schwarz, H. (1981). On the necessary interdependence between the natural sciences and theology. *Encounter*, 42, 207-223.
    (Topics):theology.

Schwarz, H. (1982). Toward a Christian stewardship of the Earth: promise and utopia. In E. R. Squiers (Eds.), *The environmental crisis: the ethical dilemma* (pp. 21-38). Mancelona, MI: Au Sable Institute of Environmental Studies.
(Topics):ethics.

Schwarz, H. (1987). Toward a Christian ecological consciousness. In F. W. Krueger (Ed.), *Christian ecology: building an environmental ethic for the twenty-first century* (pp. 12-14). North Webster, Indiana: The North American Conference on Christianity and Ecology.
(Topics):popular.

Scoby, D. R. (Ed.) (1971). *Environmental Ethics: Studies of Man's Self-Destruction*. Minneapolis: Burgess Publishing Company.
In addition to secular articles, a number of papers are included which integrate faith with environmental concerns. See especially the articles by Scoby (pp. 3, 228 & 231), Cassel (pp. 154, 225), Harris (p. 167), O'Brien (p. 197).
(Topics):ethics.

Scott, N. A. (1974). The poetry and theology of Earth: reflections on the testimony of Joseph Sittler and Gerard Manley Hopkins. *Journal of Religion*, 54(2), 102-118.
(Topics):theology.

Scoville, J. (1990) *The Au Sable Institute of Environmental Studies: Christian stewardship from an evangelical perspective*. M.A., United Theological Seminary.
(Topics):theology.

Sears, R. T. (1987). Resurrection-centered spirituality and healing the Earth. In F. W. Krueger (Ed.), *Christian ecology: building an environmental ethic for the twenty-first century* (pp. 30-31). North Webster, Indiana: The North American Conference on Christianity and Ecology.
(Topics):popular.

Sebahire, M. (1990). Saving the Earth to save life: an African point of view. *Pro Mundi Vita Studies,* 13(February), 19-26.
(Topics):African religions

Seldman, N. (1989). Social justice and solid waste management. *The Egg: A Journal of Eco-Justice,* 9(2), 2-4.
(Topics):solid waste, hazardous waste, pollution, recycling.

Sharpe, K., & Ker, J. (Eds.) (1984). *Religion and nature -- with Charles Birch and others.* Auckland, New Zealand: Univ. of Auckland Chaplaincy Publication Trust.
(Topics):popular book, theology.

Shaw, D. W. (1975). Process thought and creation. *Theology,* 78, 346-355.
(Topics):theology.

Sheaffer, J. R., & Brand, R. H. (1980). *Whatever happened to Eden?* Wheaton, IL: Tyndale House Publishers, Inc.
(Topics):book.

Sheets, J. (1971). Theological implications of ecology. *Homiletic and Pastoral Review,* 71(June), 57-66.
(Topics):theology.

Sheldon, J. K. (1989a). Twenty-one years after "the historical roots of our ecologic crisis": how has the church responded? *Perspectives on Science and Christian Faith: Journal of the American Scientific Affiliation,* 41(3), 152-158.
A study of Christian literature of the 20th century discloses a growing concern for Creation as reflected by a sharp increase in the number of publications dealing with environmental topics including a theology of creation. The publication of "The Historical Roots of Our Ecologic Crisis" by Lynn White in 1967 appears to have been a key factor contributing to the sudden appearance of a large number of creation-related publications in the early 1970's.

(Topics):history, popular.

Sheldon, J. K. (1989b). Another hot summer? The greenhouse effect, the church and you. *ESA Advocate*, 11(6), 14-15.
(Topics):popular, air pollution.

Sheldon, J. K. (1989c). How does your garden grow? a devo-
tional on creation. *ESA Advocate*, 11(9), 14-15.
(Topics):popular.

Sheldon, J. K. (1990). Creation rediscovered. *World
Christian*, 9(4), 10-19.
(Topics):species diversity, pollution, greenhouse ef-
fect, ozone, soil, land.

Sheldon, J. K. (1991). Our energy options: a response. *ESA
Advocate*, 13(6):15.
Response to article by David Gushee in the same is-
sue of *ESA Advocate*.
(Topics):energy.

Shepherd, J. B. (1970). Theology for ecology. *Catholic
World*, 211(July), 172-175.
(Topics):popular.

Sherrard, P. (1990). Confronting the ecological challenge.
Epiphany, 10(Spring), 9-18.
Also published in *Sourozh: A Journal of Orthodox
Life and Thought*, 1990, No. 41(August):7-18.
(Topics):ethics.

Sherrell, R. E. (Ed.) (1971). *Ecology: crisis and new vision.*
Richmond, VA: John Knox Press.
(Topics):book.

Sherwood, D. E., & Franklin, K. (1987). Ecology and the
church: theology and action. *The Christian Century*, 104(16),
472-474.
(Topics):popular.

Shi, D. E. (1986). *In Search of the Simple Life*. Salt Lake City, UT: Peregrine Smith Books.
(Topics):book, lifestyle.

Shinn, R. L. (1970a). Ethics and the familly of man. In M. Hamilton (Ed.), *This little planet* (pp. 127-159). New York: Charles Scribner's Sons.
(Topics):ethics.

Shinn, R. L. (1970b). Population and the dignity of man. *Christian Century*, 87(15), 442-448.
(Topics):ethics, population.

Shinn, R. L. (1972). Our technological time of troubles. *Religion in Life*, 41(4), 450-461.
(Topics):technology, popular.

Shinn, R. L. (Ed.) (1980). *Faith and science in an unjust world: report of the World Council of Churches conference on faith, science and the future. Volume 1: plenary presentations*. Philadelphia: Fortress Press.
(Topics):book.

Shinn, R. L. (1982). *Forced options: social decisions for the twenty-first century*. San Francisco: Harper and Row.
(Topics):popular, ethics.

Shinn, R. L. (1985). Eco-justice themes in Christian ethics since the 1960's. In D. Hessel (Ed.), *For Creation's sake: preaching, ecology and justice* (pp. 96-114). Philadelphia: Geneva Press.
(Topics):ethics.

Shoemaker, D. E. (1987). Loving people, loving Earth: the unity of eco-justice. *Christianity and Crisis: A Journal of Christian Opinion*, 47(August 3), 260-263.
(Topics):popular.

Shuff, K. (1989). Responsibility spurs enterprise in China. *Earthkeeping: A Quarterly on Faith and Agriculture, 5*(2), 16-17.
(Topics):appropriate technology, development.

Sickles, L. J. (1986). *A Christian view of ecology based on Genesis 1:26, 28 and 2:15*. M.A., Multnomah School of the Bible, Portland, Oregon.
(Topics):theology.

Sider, R. J. (1977, June). Sharing wealth: the church as the Biblical model for public policy. *Christian Century*, pp. 560-565.
(Topics):lifestyle.

Sider, R. J. (Ed.) (1980). *Living more simply: biblical principles & practical models*. Downers Grove, IL: Inter-Varsity Press.
(Topics):lifestyle.

Sider, R. J. (1982). *Lifestyle in the eighties: an evangelical commitment to simple lifestyle*. Louisville, KY: Westminster/John Knox.
(Topics):lifestyle.

Sider, R. J. (1988, November/December). The Limitations of politics. *ESA Advocate*, pp. 1-7.
(Topics):popular.

Sider, R. J. (1990a). Justice, peace, & the integrity of creation. *World Christian, 9*(5), 27-30.
Report on the World Council of Churches assembly at Seoul, Korea, in 1990 on the topic of JPIC (Justice, Peace, and the Integrity of Creation).
(Topics):popular, ecojustice.

Sider, R. J. (1990b). *Rich Christians in an age of hunger* (3rd ed.). Downers Grove, IL: Inter-Varsity Press.
(Topics):lifestyle, population.

Sider, R. J. (1990c). Buses and bicycles. *ESA Advocate*, 12(3):1-3.
(Topics):energy, pollution.

Sider, R. J. (1991). Green theology. *ESA Advocate*, 13(6):1-4.
Briefly reviews the joint meeting on June 2-3, 1991, of religious, political, and scientific leaders.
(Topics):theology, popular.

Sider, R. J., & Taylor, R. (1982). *Nuclear holocaust and Christian hope*. Downers Grove, IL: Inter-Varsity Press.
(Topics):nuclear.

Sikkema, S. (1987). The global farm crisis: European and North American Christians confer. *Earthkeeping: A Quarterly on Faith and Agriculture*, 3(5), 12-15.
(Topics):land, agriculture.

Sikkema, S. (1989). Struggling with two-thirds world obligations and GATT. *Earthkeeping: A Quarterly on Faith and Agriculture*, 5(1), 17.
(Topics):popular, agriculture, land.

Simmons, H. C. (1990). Religious education and the integrity of creation. *Religious Education*, 85(Winter), 5-50.
(Topics):education.

Simmons, P. (Ed.) (1980). *Issues in Christian ethics*. Nashville, TN: Broadman Press.
(Topics):ethics.

Simon, J. L., & Kahn, H. (1984). *The resourceful Earth*. New York: Basil Blackwell.
A secular book written by economists based on a cornucopian world view that minimizes present environmental problems.
(Topics):cornucopian, economics.

Sims, B. J. (1990). The North American Conference on Religion and Ecology, Washington, DC, May 16-18, 1990: a

report to the presiding Bishop. *Anglican and Episcopal History*, 59(December), 441-452.
(Topics):popular.

Sine, T. (1981). *The mustard seed conspiracy*. Waco, TX: Word Books.
(Topics):lifestyle.

Sire, J. W. (1976). *The universe next door: a basic world view catalog*. Downers Grove, IL: Inter-Varsity Press.
(Topics):world view.

Sisk, J. P. (1984). Call of the wild. *Commentary*, 78(1), 52-55.
(Topics):popular.

Sittler, J. (1954). A theology for Earth. *The Christian Scholar*, 37(3), 367-374.
(Topics):theology.

Sittler, J. (1962). Called to unity. *The Ecumenical Review*, 14(2), 175-187.
(Topics):history, theology.

Sittler, J. (1964a). *The care of the Earth*. Philadelphia: Fortress Press.
(Topics):book, theology.

Sittler, J. (1964b). Nature and grace: reflections on an old rubric. *Dialog*, 3, 252-256.
(Topics):theology.

Sittler, J. (1970a). Ecological commitment as theological responsibility. *Zygon*, 5(June), 172-181.
(Topics):theology.

Sittler, J. (1970b). *The ecology of faith*. Philadelphia: Fortress Press.
(Topics):theology.

Sittler, J. (1971). Ecological commitment as theological responsibility. *Southwestern Journal of Theology*, 13(Spring), 35-45.
(Topics):theology.

Sittler, J. (1972a). *Essays on nature and grace.* Philadelphia: Fortress Press.
(Topics):theology.

Sittler, J. (1972b). Scope of Christological reflection. *Interpretation: A Journal of Bible and Theology*, 26(July), 328-337.
(Topics):theology.

Sittler, J. (1981). *Grace notes and other fragments.* Philadelphia: Fortress Press.
A short book of essays, several of which relate to the environment.
(Topics):popular book.

Sittler, J. (1986). *Gravity and grace.* Minneapolis: Augsburg Publishing House.
(Topics):popular book.

Skillin, J. W. (1982). Ethics and justice: what should governments do for the environment? In E. R. Squiers (Ed.), *The environmental crisis: the ethical dilemma* (pp. 227-236). Mancelona, MI: Au Sable Institute of Environmental Studies.
(Topics):ethics.

Skoglund, J. E. (1974). Ecology and justice [special issue]. *Foundations*, 17(April-June), 99-172.
Thematic issue dealing with ecojustice, edited by J. E. Skoglund.
(Topics):popular, ecojustice.

Skolimowski, H. (1983). Eco-ethics as the imperative of our times. *Epiphany*, 3(3), 26-34.
(Topics):popular.

Skolimowski, H. (1985). *Eco-theology: toward a religion for our times*. Madras, India: Vasawata Press.
(Topics):theology.

Slattery, P. (1991). *Caretakers of creation: farmers reflect on their faith and work*. Minneapolis: Augsburg.
(Topics):agriculture, land, popular book.

Smalley, W. A. (1962). The gospel, the church, and the population explosion. *Journal of the American Scientific Affiliation*, 14(March), 11-14.
(Topics):population.

Smith, A. J. (1990). Stemming the tide of our own destruction. *United Evangelical Action*, 49(May), 10-11.
(Topics):recycling, waste.

Smith, H. (1972). Tao now: an ecological testament. In I. Barbour (Ed.), *Earth might be fair*. Englewood Cliffs, N.J.: Prentice-Hall, Inc.
Suggests that a more pantheistic religion is needed to meet the environmental crisis.
(Topics):popular.

Smith, H. (1987). Animal rights and Christian tradition. In F. W. Krueger (Ed.), *Christian ecology: building an environmental ethic for the twenty-first century* (p. 53). North Webster, Indiana: The North American Conference on Christianity and Ecology.
(Topics):animal rights.

Smith, H. B. (1988). This old hothouse. *Christianity Today*, 32(15), 15.
(Topics):popular.

Smith, H. L. (1970). Religious and moral aspects of population control. *Religion Life*, 39, 193-204.
(Topics):population.

Smith, I. (1987). Christian ecological economics: part two: intrinsic values. In F. W. Krueger (Ed.), *Christian ecology: building an environmental ethic for the twenty-first century* (p. 71). North Webster, Indiana: The North American Conference on Christianity and Ecology.
(Topics):economics.

Smith, M. (1987). "Shalom": Christian stewardship as applied to issues. In F. W. Krueger (Ed.), *Christian ecology: building an environmental ethic for the twenty-first century* (p. 66). North Webster, Indiana: The North American Conference on Christianity and Ecology.
(Topics):popular.

Smith, R. L. (1971). Old Testament concepts of stewardship. *Southwestern Journal of Theology, 21*(3), 13.
(Topics):theology, stewardship.

Snell, P. (1979). The Bible and ecology. *Bible Today, 104*, 2180-2185.
(Topics):popular.

Snyder, G. (1969). *Earth household.* New York: New Directions.
Has a pantheistic focus.
(Topics):popular.

Snyder, G. (1980). Where we are. *Parabola: The Magazine of Myth and Tradition, 5*(1), 81-85.
Has a pantheistic focus.
(Topics):popular.

Snyder, H. A. (1982). *Liberating the church: the ecology of the church and kingdom.* Downers Grove, Ill.: Inter-Varsity Press.
(Topics):book.

Snyder, H. A. (1983). Kingdom ecology: a model for the church in the world. *TSF Bulletin, 6*(4), 4-7.
(Topics):popular.

Snyder, H. A. (1985). *A kingdom manifesto: calling the Church to live under God's rein.* Downers Grove, Ill.: Inter-Varsity Press.
(Topics):book.

Soelle, D. (1984). *To work and to love: a theology of creation.* Philadelphia: Fortress Press.
(Topics):theology.

Soleri, P. (1981). *The omega seed: an eschatological hypothesis.* Garden City, N.Y.: Anchor Books, Doubleday.
(Topics):book.

Somplatsky-Jarman, W. (1990). For a more inclusive environmental agenda. *Christianity and Crisis,* 50(7):144-145.
Notes the lack of involvement of minority groups in major environmental organizations and the siting of toxic waste sites in poor and ethnic neighborhoods
(Topics):ecojustice, racial justice.

Sorenson, J. (1986). Living on the land: imaging our ways. *Word & World,* 6(1), 29-34.
(Topics):land.

Sorokin, V., & Vladimir, I. (1985a). Hope, unity and peace: vital concerns of the conference of European churches, part I. *The Journal of the Moscow Partriarchate,* 9, 94-96.
(Topics):popular.

Sorokin, V., & Vladimir, I. (1985b). Hope, unity and peace: vital concerns of the conference of European churches, part II. *The Journal of the Moscow Partriarchate,* 10, 55-60.
(Topics):popular.

Spaling, H. H. (1982). Land and life: the threatened link. *Journal of the American Scientific Affiliation,* 34(4), 212-218.
(Topics):land.

Spradley, J. (1985). A Christian view of the physical world. In A. Homes (Ed.), *The making of a Christian mind: a Christian world view and the academic enterprise.* Downers Grove, IL: Inter-Varsity Press.
(Topics):history.

Spradley, J. L. (1970). Christian roots of science. *Christianity Today,* 14(12), 519-520.
(Topics):popular.

Spring, D., & Spring, E. (1974). *Ecology and religion in history.* New York: Harper & Row Publishers.
(Topics):book, history.

Squadrito, K. M. (1979). Locke's view of dominion. *Environmental Ethics,* 1(3), 255-262.
(Topics):history, ethics.

Squiers, E. R. (Ed.) (1982). *The environmental crisis: the ethical dilemma.* Mancelona, MI: Au Sable Institute of Environmental Studies.
(Topics):ethics, book.

Squiers, E. R. (1989). Wealth and waste and writing on the wall: (or, somebody go get Daniel). *Perspectives on Science and Christian Faith,* 41(June), 69-75.
(Topics):energy,resources, biodiversity, pollution.

Squiers, E. R. (1990). Good news bad news re: our environment. *United Evangelical Action,* 49(3), 4-9.
(Topics):popular.

St. John, D. (1982). Ecological prayer: toward an ecological spirituality. *Encounter,* 43, 337-348.
(Topics):popular.

St. John, D. P. (1987). Creation and the churches: a call for a new reformation. In F. W. Krueger (Ed.), *Christian ecology: building an environmental ethic for the twenty-first*

century (p. 50). North Webster, Indiana: The North American Conference on Christianity and Ecology.
(Topics):popular.

Stacey, W. D. (1956). Christian view of nature. *Expository Times*, 67, 364-367.
(Topics):theology.

Stafford, T. (1990). Animal lib. *Christianity Today*, 34(9), 19-23.
(Topics):animal rights.

Stamm, B. J. (1987). Little liberation in rural exodus. *Earthkeeping: A Quarterly on Faith and Agriculture*, 3(3), 23.
(Topics):land.

Stange, D. C. (1971). Bibliography on religion and ecology for Lutherans. *Lutheran Quarterly*, 23(November), 329-334.
(Topics):theology.

Statt, J., & Coote, R. (Eds.) (1980). *Down to Earth: studies in Christianity and culture*. Grand Rapids, MI: Eerdmans Publishing Co.
See especially pages 97-114.
(Topics):popular book.

Steck, O. H. (1980). *World and environment. Biblical encounters series*. Nashville, TN: Abingdon Press.
Extensive bibliography of German citations.
(Topics):theology.

Steele, T. (1987). What is the proper relation of humanity to nature? *The AME Zion Quarterly Review*, 97, 38-41.
(Topics):popular.

Steffenson, D.,Cook, R. S., & Herrscher, W. J. (Eds.) (1973). *Ethics for environment: three religious strategies*. Green Bay: University of Wisconsin.
Conference proceedings.

(Topics):ethics.

Stefferud, A. (Ed.) (1972). *Christians and the good Earth.*
New York: Friendship Press.
>Publication from the Faith-Man-Nature group.
>(Topics):book.

Stein, G. J. (1981). The "new" ethics of survival. *Saint Luke's Journal of Theology,* 24(3), 200-216.
>Addresses the implications for Christianity of the new "lifeboat" ethics, but not from an overtly Christian position.
>(Topics):ethics.

Steindl-Rast, D. (1983). Nature and the Poetic Intuition. *Epiphany,* 3(3), 7-13.
>(Topics):popular.

Stevenson, W. T. (1976). Historical consciousness and ecological crisis: a theological perspective. *Anglican Theological Review Supplementary Series,* 7(November), 99-111.
>(Topics):history, theology.

Steward, J. (1986, July-September). Water of life. *Together: A Journal of World Vision International,* p. 23-24.
>(Topics):water.

Steward, J. (1989, January-March). A sabbath rest for your land? *Together: A Publication of World Vision,* p. 15.
>(Topics):land.

Steward, R. G. (1990). *Environmental Stewardship.*
Downers Grove, IL.: Inter-Varsity press.
>Part of the Global Issues Bible Studies.
>(Topics):educational, contains six Bible studies.

Stewart, C. (1983). *Nature in grace: a study in the theology of nature.* Macon, GA: Mercer University Press.

NABPR Dissertation Series, No. 3. Abstract published in the *Harvard Theololgical Review*, 74: 409, Oct. 1981.
(Topics):theology.

Stewart-Kroeker, C., & Kroeker, P. T. (1989). Farm crisis has natural and spiritual explanations. *Earthkeeping: A Quarterly on Faith and Agriculture*, 5(4/5), 31.
(Topics):agriculture, land.

Stipe, C. E. (1970). Some theological issues. *Journal of the American Scientific Affiliation*, 22(2), 47-48.
(Topics):population.

Stipe, C. E. (1974). Suffers from inadequate preparation. *Journal of the American Scientific Affiliation*, 26(1), 21.
(Topics):population.

Stivers, R. L. (1976). *The sustainable society*. Philadelphia: Westminster Press.
(Topics):economics.

Stivers, R. L. (1979). The sustainable society: religious and social implications. *Review of Religious Research*, 21(1), 71-86.
(Topics):economics, popular.

Stivers, R. L. (1984). *Hunger, technology & limits to growth: Christian responsibility for three ethical issues*. Minneapolis: Augsburg Publishing House.
(Topics):agriculture, ethics, book.

Stivers, R. L. (1986). Justice, participation, and sustainable sufficiency. In B. W. Harrision, R. L. Stivers, & R. H. Stone (Eds.), *The public vocation of Christian ethics* (pp. 179-191). New York: The Pilgrim Press.
(Topics):ethics, population.

Stone, G. C. (Ed.) (1971). *A new ethic for a new Earth*. New York: Friendship Press.

From the Faith-Man-Nature group.
(Topics):popular book.

Stone, P. (1989, January/February). Environmentalism and spirituality part II: Christian ecology. *Mother Earth News*, pp. 58-61.
(Topics):popular.

Stone, R. H. (1972). Ethics and growth. *South East Asia Journal of Theology*, 14(1), 40-62.
(Topics):ethics.

Stoner, J. K. (1987). War, conflict and environmental degradation: part two: a theological foundation. In F. W. Krueger (Ed.), *Christian ecology: building an environmental ethic for the twenty-first century* (p. 75). North Webster, Indiana: The North American Conference on Christianity and Ecology.
(Topics):war.

Storck, T. (1985). Living with nature. *Social Justice Review*, 76(May/June), 78-79+.
(Topics):popular.

Stott, J. (1984). *Involvement: being a responsible Christian in a non-Christian society.* Old Tappan, New Jersey: Fleming H. Revell Company.
See chapter six, "Our Human Environment."
(Topics):popular book.

Strachan, G. (1985). *Christ and the cosmos.* Dunbar, Scotland: Labarum Publications.
(Topics):popular book.

Stratman, T. B. (1982). St. Francis of Assisi: brother to all creatures. *Spirituality Today*, 34, 222-232.
Examines the beliefs of St. Francis of Assisi and suggests that we would be wise to emulate him.
(Topics):history.

Strauss, J. D. (1974) *A theology of nature and the ecological crisis.* D.Min., Eden Theological Seminary.
(Topics):theology.

Streiffert, K. G. (1989a). The earth groans, and Christians are listening. *Christianity Today,* 33(13), 38-41.
(Topics):popular.

Streiffert, K. G. (1989b, July/August). God bless our modern ark. *Christian Outdoorsman,* pp. 10-11.
(Topics):popular.

Streiffert, K. G. (1989c, March/April). Paradise--preserved. *Christian Outdoorsman,* p. 13.
(Topics):popular.

Stuhlmacher, P. (1987). The ecological crisis as a challenge for Biblical theology. *Ex Auditu: An Annual of the Frederick Neumann Symposium on Theological Interpretation of Scripture,* 3, 1-15.
Translated by J. M. Scott.
(Topics):theology.

Suhor, M. L. (Ed.) (1990). Creating a new heaven and a new earth. *The Witness,* 73(September), 6-23.
(Topics):popular.

Suld, H. B. D. (1962) *The relation of the doctrine of creation to the person of Jesus Christ in New Testament theology.* Ph.D., McGill University (Canada).
(Topics):theology.

Sullivan, E. (1970). The Growing Danger to Our Environment. *Sign,* 49, 12-23.
(Topics):popular.

Swartley, W. M. (1978). Biblical sources of stewardship. In M. E. Jegen & B. V. Manno (Eds.), *The Earth is the Lord's: essays on stewardship* (pp. 22-43). New York: Paulist Press.
(Topics):popular, stewardship.

Sweeting, G. (1980). Entering the twilight age: the energy problem comes full circle, exposing our sin and greed. *Christianity Today*, 24(June 27), 28-29.
(Topics):energy.

Sweetland, G. B. (1975). Salvation of the cosmos in Eastern Christian tradition. *Journal of Ecumenical Studies*, 12(2), 253-256.
(Topics):theology, popular.

Swift, D. L. (1982). Ethics in environmental policy and standards. In E. R. Squiers (Ed.), *The environmental crisis: the ethical dilemma* (pp. 235-244). Mancelona, MI: Au Sable Institute of Environmental Studies.
(Topics):ethics.

Swincer, G. (1989, January - March). Soil conservation: structure and nutrients. *Together, A Publication of World Vision*, p. 14.
(Topics):land, soil conservation.

Sytsma, L. (1991). Hazardous waste: a problem that can't be buried. *ESA Advocate*, 13(6):8-11.
(Topics):hazardous waste, nuclear.

Tanenbaum, M. C. (1985). Earth day 2000 years ago. In N. C. Murphy (Ed.), *Teaching and preaching stewardship: an anthology* (pp. 217-222). New York: Commission on Stewardship, National Council of the Churches of Christ in the U.S.A.
(Topics):popular.

Tang, E. (1990). What has religion to do with ecology? *Pro Mundi Vita Studies*, 13(February), 6-18.
(Topics):popular, Hindu, Taoism, Confucianism, Buddhism, Eastern religions.

Tanner, K. (1988). *God and creation in Christian theology: tyranny or empowerment?* Oxford: Basil Blackwell, Ltd.

(Topics):theology.

Tatman, L. (1990a). Hildegard of Bingen. *Daughters of Sarah*, 16(3), 3.
  (Topics):ecofeminism.

Tatman, L. (1990b). She cried. *Daughters of Sarah*, 16(3),22-23.
  (Topics):ecofeminism.

Taylor, G. R. (1970). *The doomsday book: can the world survive?* New York: World.
  (Topics):popular book.

Taylor, J. (1990). Soviet Muslims: ecological crisis, *glasnost* and the gospel. *World Christian*, 9(May), 10-17.
  (Topics):popular.

Taylor, J. V. (1977). *Enough is enough.* Minneapolis: Augsburg.
  (Topics):lifestyle.

Taylor, L. H. (1958). *The new creation.* New York: Pageant Press.
  (Topics):theology.

Taylor, R. (1982). *A community of stewards.* Minneapolis: Augsberg.
  (Topics):book.

Teilhard de Chardin, P. (1956). *Man's place in nature.* New York: Harper.
  (Topics):book.

Teilhard de Chardin, P. (1959). *The phenomenon of man.* New York: Harper.
  (Topics):book.

Teilhard de Chardin, P. (1964). *The future of man.* New York: Harper and Row.

(Topics):book.

Teitelbaum, B. (1985). The return voyage: fighting for home and family. *Epiphany*, 6(2), 70-76.
(Topics):popular.

Temple, W. (1934). *Nature, Men and God*. London: Macmillan.
(Topics):theology.

Terman, C. R. (1974). Sociobiology and population problems: perspectives. *Journal of the American Scientific Affiliation*, 26(1), 6-13.
(Topics):population.

Terman, C. R. (1982). The kepone problem in the James River and adjacent areas of Virginia: a classic example for ethical and scientific consideration. In E. R. Squiers (Ed.), *The environmental crisis: the ethical dilemma* (pp. 245-260). Mancelona, MI: Au Sable Institute of Environmental Studies.
(Topics):ethics, pollution, hazardous waste, toxic waste.

Terman, M. R. (1982). Ethical and ecological basis for earth sheltered housing. In E. R. Squiers (Ed.), *The environmental crisis: the ethical dilemma* (pp. 309-314). Mancelona, MI: Au Sable Institute of Environmental Studies.
(Topics):energy, ethics.

Teske, M. (1971). On recycling symbols in dialogue between theologians and scientists. *The Lutheran Quarterly*, 23(4), 317-328.
(Topics):theology.

Testeman, D. E. (1991). Missionary earthkeeping: glimpses of the past, visions of the future. In G. T. Prance & C. B. DeWitt (Eds.), *Missionary Earthkeeping*. Macon, Georgia: Mercer University Press.
(Topics):popular.

Thavis, J. (1988, Nov. 25). Pope, Vatican science academy raise flag warning that ecology concerns are serious moral issue. *National Catholic Reporter*, p. 8.
(Topics):popular.

Thomas, C. (1988). Romans 8: a challenge to Christian involvement. *Trinity Seminary Review*, 10(1):31-37.
(Topics):theology.

Thomas, J. M. (1987). *Ethics and Technoculture*. Lanham, Maryland: University Press of America.
(Topics):ethics, technology.

Thomas, J. M. (1991). Introduction. In G. T. Prance & C. B. DeWitt (Eds.), *Missionary Earthkeeping*. Macon, Georgia: Mercer University Press.
(Topics):popular.

Thomas, K. (1983). *Man and the natural world: a history of the modern sensibility*. New York: Pantheon Books.
(Topics):history.

Thomas, K. (1984). *Man and the natural world -- changing attitudes in England 1500-1800*. Harmondsworth, England: Penguin Books.
(Topics):history.

Thompson, T. K. (Ed.) (1960). *Stewardship in contemporary theology*. New York: Association Press.
(Topics):stewardship.

Thompson, T. K. (1965). *Handbook of stewardship procedures*. Englewood Cliffs, N. J.: Prentice-Hall, Inc.
(Topics):stewardship.

Thompson, W. M. (1987). "Dappled and Deep Down Things": A Meditation on Christian Ecological Trends. *Horizons*, 14(Spring), 64-81.
(Topics):theology.

Thorkelson, W. L. (1985). Rural life ministers urge respect for environment, new politics, economies for land and resources. *National Catholic Reporter*, 21(September 27), 26-27.
(Topics):popular.

Thorpe, W. H., Barker, E., Canon Dillistone, & Canon Edward Carpenter (1970). Nature. *The Modern Churchmen*, 14(October), 9-42.
Transcript of a panel discussion including section on "Man's Stewardship towards Nature." Includes contributions by W. H. Thorpe, Edwin Barker, Canon Dillistone and Canon Edward Carpenter.
(Topics):popular.

Thurber, L. N. (1990). Care for the creation as mission responsibility. *International Review of Mission*, 79(April), 143-149.
(Topics):popular.

Tiemstra, J. P. (1977). Energy and Christian stewardship: an economist's appraisal. *Journal of the American Scientific Affiliation*, 29(3), 99-102.
(Topics):energy, economics.

Tiemstra, J. P. (1981). Economic causes of soil erosion in the United States. *Journal of the American Scientific Affiliation*, 33(3), 175-178.
(Topics):land, economics, resources.

Tilak, S. J. (1977) *Nature, history and spirit in trialogue among western secularism, Hindu spiritualism, and Christian trinitarian faith.* S.T.D., Lutheran School of Theology at Chicago.
(Topics):theology.

Tillich, P. (1988). *The spiritual situation in our technical society.* Macon, Georgia: Mercer University Press.
J. M. Thomas served as the editor of this series of works by Tillich.

(Topics):theology, technology.

Timm, R. E. (1986). Let's not miss the theology of the creation accounts. *Currents in Theology and Mission*, 13(2), 97-105.
(Topics):theology, popular.

Timm, R. E. (1990). Divine majesty, human vicegerency, and the fate of the earth in early Islam. *Hamdard Islamicus*, 13(Spring), 47-57.
(Topics):Islam, popular.

Tinker, G. E. (1986). Native Americans and the land: "the end of living, and the beginning of survival." *Word & World*, 6(1), 66-74.
(Topics):land.

Tischel, P. (1983). The return to unity: meditations on the destiny of man. *Epiphany*, 3(4), 76-81.
(Topics):popular.

Todd, J. (1983). Earth and stewardship: a conversation with John Todd. *Epiphany*, 3(4), 38-46.
(Topics):popular.

Toerien, D. F. (1987). Water, the limiting resource. In W. S. Vorster (Ed.), *Are We Killing God's Earth?* (pp. 68-79). Pretoria, South Africa: University of South Africa.
(Topics):water, pollution.

Tokar, B. (1987). *The green alternative: creating an ecological future*. San Pedro: R. & E. Miles.
(Topics):land, agriculture.

Tokuzen, Y. (1983). Nature and the natural in Luther's thought. *East Asia Journal of Theology*, 1(1), 31-37.
(Topics):theology.

Tomlin, E. W. F. (1985). Ecological reflection. *Heythrop Journal*, 26, 187-196.

        Reviews several books on nature and man's responsibility.
        (Topics):popular.

Toulmin, S. (1985). Nature and nature's God. *Journal of Religious Ethics*, 13(Spring), 37-52.
        (Topics):theology.

Toulmin, S. (1987). Religion and the idea of nature. In F. T. Birtel (Ed.), *Religion, Science, and Public Policy* (pp. 67-78). New York: The Crossroad Publishing Company.
        (Topics):history.

Toynbee, A. (1972). The religious background of the present environmental crisis. *International Journal of Environmental Studies*, 3, 141-146.
        Supports White's 1967 argument that Judeo-Christian tradition is responsible for the ecological crisis. Also appears in Spring and Spring, eds. 1974.
        (Topics):popular.

Toynbee, A. (1973). The genesis of pollution. *Horizon*, 15(3), 4-9.
        Supports White's 1967 argument that Judeo-Christian tradition is responsible for the ecological crisis.
        (Topics):popular.

Treece, A. (1983). St. Gall -- an ancient plan for modern self-sufficiency. *Epiphany*, 3(3), 80-87.
        (Topics):lifestyle.

Trible, P. (1971). Ancient priests and modern polluters. *Andover Newton Quarterly*, 12, 74-79.
        Also published in *Foundations*, 17(2):158-163.
        (Topics):theology.

Trickett, D. G. (1982). *Toward a Christian theology of nature: a study based on the thought of H. Richard Niebuhr.* Ph.D., Southern Methodist University.
        (Topics):theology.

Trotter, K. (1987). Spiritualities in environmental education: part one: redemptive evocative teaching. In F. W.
Krueger (Ed.), *Christian ecology: building an environmental ethic for the twenty-first century* (p. 80). North Webster,
Indiana: The North American Conference on Christianity
and Ecology.
(Topics):education.

Tuan, Y. (1968). Discrepancies between environmental attitude and behavior: examples from Europe and China. *The
Canadian Geographer*, 12(3), 176-191.
Discusses the environmental damage in countries
not influenced by the Judeo-Christian tradition as an argument against White 1967. Reprinted in Spring and Spring,
eds. 1974.
(Topics):popular, history.

Tuan, Y. (1970). Our treatment of the environment in ideal
and actuality. *American Scientist*, 58(3), 244-249.
Discusses the impact of man on nature in both the
Christian west as well as in the non-Christian world in support of the position that environmental degradation has oc-
cured historically in cultures not significantly influenced
by the Judeo-Christian tradition.
(Topics):popular.

Tuck, W. P. (1972). Church and ecological action. *Review
and Expositor: A Baptist Theological Journal*, 69(1), 67-76.
(Topics):theology, popular.

Turner, E. S. (1964). *All heaven in a rage*. New York: St.
Martin's.
(Topics):animal rights.

Umphrey, L. (1971). Pitfalls and promises of Biblical texts
as a basis for a theology of nature. In G. C. Stone (Ed.), *A
new ethic for a new Earth*. New York: Friendship Press.
(Topics):theology.

Underwood, R. A. (1969). Ecological and psychedelic approaches to theology. *Soundings*, 52(Winter), 365-393.
    Suggests that a new kind of mystical religion is needed to meet the ecological crisis rather than one based on the Judeo-Christian tradition.
    (Topics):popular.

United Methodist Church (General Conference) (1984). *Faithful witness on today's issues: environmental stewardship*. Nashville, TN: Discipleship Resources for Church and Society.
    15-page booklet. Available from Discipleship Resources for Church and Society, 1908 Grand Ave., P. O. Box 189, Nashville, TN 37202. Order No. CS95.
    (Topics):popular.

United Presbyterian Church in the U.S.A. (182nd General Assembly) (1970). Environmental tragedy. *Church and Society*, 61(September/October), 21-29.
    (Topics):popular.

United Presbyterian Church in the U.S.A. (183rd General Assembly) (1971). Statement on environmental renewal. *Church and Society*, 61(July/August), 21-23.
    (Topics):popular.

United Presbyterian Church in the U.S.A. (185th General Assembly) (1973). The global population crisis. *Church and Society*, 64(November/December), 31-35.
    (Topics):population.

United Presbyterian Church in the U.S.A. (186th General Assembly) (1974). Christian responsibility in the energy crunch. *Church and Society*, 65(September/October), 47-55.
    (Topics):energy.

United Presbyterian Church in the U.S.A. (188th General Assembly) (1976). Economic justice within environmental limits: the need for a new economic ethic. *Church and Society*, 67(September/October), 5-56.

(Topics):economics.

United Presbyterian Church in the U.S.A. (1982). *The Theology of Stewardship* (17 pp.); and *A Study Guide: The Theology of Stewardship* (18 pp.) (by Robert Case). New York: Advisory Council on Discipleship and Worship.
Available from Advisory Council on Discipleship and Worship, 475 Riverside, Room 1020, New York, NY 10115.
(Topics):stewardship, theology.

Vanackere, H. (Ed.) (1990). Christians and the ecological consciousness. *Pro Mundi Vita Studies*, 13(February), 6-43.
(Topics):popular.

Vander Zee, D. (1985). The environmental impact of being busy in the Creation. *Pro Rege*, 13(3):12-20.
Pro Rege is published quarterly by the faculty of Dordt College, Sioux Center, Iowa.
(Topic):popular.

Vander Zee, D. (1987). Gift or grasp. *Christian Educators Journal*, 27(2), 15-17.
(Topics):education.

Vander Zee, D., and Vos, R. (1990). Trends in agriculture: sustainability. *Pro Rege*, 18(3):19-28.
Pro Rege is published quarterly by the faculty of Dordt College, Sioux Center, Iowa.
(Topic):agriculture.

van Donkersgoed, E. (1986). Erosion: economic or ethical challenge. *Earthkeeping: A Quarterly on Faith and Agriculture*, 2(2), 16.
(Topics):land.

van Donkersgoed, E. (1989). Decoupling: a family farm and stewardship approach. *Earthkeeping: A Quarterly on Faith and Agriculture*, 5(4/5), 12-15.
(Topics):land, agriculture.

van Donkersgoed, E. (1990). A vision to be ignored: a faith based agenda for the 1990s for Christian farmers in Canada. *Earthkeeping: A quarterly on faith and agriculture*, 6(1):13-15.
(Topics):land, agriculture.

van Drimmelen, R. (1990). A prayer and a task. *One World*, 161(December), 18-19.
Report on the World Council of Church's assembly in Kuala Lampur in preparation for the WCC's Seventh Assembly in Canberra.
(Topics):popular.

Van Dyke, F. G. (1985). Beyond sand country: a Biblical perspective on environmental ethics. *Journal of the American Scientific Affiliation*, 37(1), 40-48.
(Topics):ethics.

Van Dyke, F. G. (1986). Annotated bibliography on planetheonomics. In *1986 Au Sable forum on planetheonomics*. Au Sable Institute of Environmental Studies, 7526 Sunset Trail N.E., Mancelona, Michigan 49659. Available by writing to Au Sable.
(Topics):economics, bibliography.

Van Dyke, F. G. (1988). Planetary economies and ecologies: the Christian world view and recent literature. *Perspectives on Science and Christian Faith*, 40(2), 66-71.
(Topics):economics.

Van Eeden, I. J. (1987). Ethical questions pertaining to the 'soft explosion.' In W. S. Vorster (Ed.), *Are We Killing God's Earth?* (pp. 80-88). Pretoria, South Africa: University of South Africa.
(Topics):ethics, population, resources.

Van Elderen, M. J. (1987). Taking another look at our relationship to nature. *One World*, 122, 24.
(Topics):popular.

Van Geest, W. (1990). The whole plan of salvation. *Faith Today*, 8(Mr/Ap), 37.
  (Topics):popular.

Van Gerwen, J. (1990). Ecology and Christian vision: some conclusions. *Pro Mundi Vita Studies*, 13(February), 42-43.
  (Topics):popular.

van Hoeven, J. W. (1989a). The church and the drugging of creation. *Perspectives: A Journal of Reformed Thought*, 4(2), 3.
  (Topics):stewardship, world view, popular.

van Hoeven, J. W. (Ed.) (1989b). *Justice, peace and the integrity of creation.*
  Papers and Bible studies edited by van Hoeven for the World Alliance of Reformed Churches Assembly, Seoul, Korea, August 1989. Available at WARC Office, 150 route de Ferney, 1211 Geneva 2.
  (Topics): popular.

Van Leeuwen, R. C. (1989). Enjoying creation -- within limits. *Christianity Today*, 33(8), 34-37.
  (Topics):popular, lifestyle.

Van Leeuwen, R. C. (1991). Resurrection and the vindication of creation. In C. B. DeWitt (Ed.), *The environment and the Christian: what does the New Testament say about the environment?* Grand Rapids, Michigan: Baker Book House.
  (Topics):theology.

Van Til, C. (1946). Nature and scripture. In N. B. Stonehouse & P. Wooley (Eds.), *The infallible word.* Grand Rapids, MI: Baker Book House.
  (Topics):theology.

Vander Schaaf, J. (1987). Holistic action required for prairie agriculture. *Earthkeeping: A Quarterly on Faith and Agriculture*, 3(5), 4-6.
(Topics):land, agriculture.

Vanstone, W. H. (1977). On the being of nature. *Theology*, 80(July), 279-283.
(Topics):popular, theology.

Vaux, K. (1970). *Subduing the cosmos: cybernetics and man's future*. Richmond, VA: John Knox Press.
(Topics):theology.

Veeraraj, A. (1990). God is green. *International Review of Mission*, 79(April), 187-192.
(Topics):popular.

Verghese, P. (1974). Mastery and mystery. *Religion and Society*, 21(December), 29-38.
(Topics):theology.

Verhey, A. (1985). The morality of genetic engineering. *Christian Scholar's Review*, 14(2), 124-139.
(Topics):ethics, genetic engineering.

Vickery, J. (1979). God in America's wilderness consciousness. *Spiritual Life*, 25, 47-54.
(Topics):history, wilderness.

Visick, V. (1991). Creation's care and keeping in the life of Jesus. In C. B. DeWitt (Ed.), *The environment and the Christian: what does the New Testament say about the environment?* Grand Rapids, Michigan: Baker Book House.
(Topics):theology.

Visser, C. (1989). Recognizing environmental threats and security. *Earthkeeping: A Quarterly on Faith and Agriculture*, 5(1), 21-22.
(Topics):popular, land.

von Rad, G. (1962). *Old Testament theology, Vol. 1* (Stalker, D. M. G., Trans.). New York: Harper & Row, Publishers.
(Topics):theology.

von Rad, G. (1965). *Old Testament theology, Vol. 2* (Stalker, D. M. G., Trans.). New York: Harper & Row, Publishers.
(Topics):theology.

von Rad, G. (1966). *The problem of the Hexateuch and other essays* (Trueman Dicken, E. W., Trans.). New York: McGraw-Hill.
(Topics):theology.

von Rad, G. (1972). *Genesis.* Philadelphia: Westminster Press.
(Topics):theology.

von Rohr Sauer, A. (1974). Ecological notes from the Old Testament. In H. N. Bream (Ed.), *A light unto my path* (pp. 421-434). Philadelphia: Temple University Press.
(Topics):popular, theology.

Vorspan, A. (1970). The crisis of ecology: Judaism and the environment. In A. Vorspan (Ed.), *Jewish values and social crisis* (pp. 179-198). New York: Union of American Hebrew Congregations.
(Topics):Judaism.

Voth, E. H. (1982). Time in a Christian environmental ethic. In E. R. Squiers (Ed.), *The environmental crisis: the ethical dilemma* (pp. 57-66). Mancelona, MI: Au Sable Institute of Environmental Studies.
(Topics):ethics.

Wade, C. R. (1990). Good, but broken. *Southwestern Journal of Theology,* 32(Spring), 6-9.
(Topics):theology.

Walker, L. C. (1980). Ecologic concepts in forest management. *Journal of the American Scientific Affiliation*, 32(4), 207-214.
(Topics):resources.

Walker, L. C. (1986). Resource managers and the environmental ethic. *Journal of the American Scientific Affiliation*, 38(2), 96-102.
(Topics):resources.

Wallace-Hadrill, D. S. (1968). *The Greek patristic view of nature*. Manchester: Manchester University Press.
(Topics):history.

Walsh, B. J. (1987). Theology of hope and the doctrine of creation: an appraisal of Jurgen Moltmann. *The Evangelical Quarterly*, 59, 53-76.
(Topics):theology.

Walsh, B. J., & Middleton, J. R. (1984). *The transforming vision: shaping a Christian world view*. Downers Grove, IL: Inter-Varsity Press.
    Includes a discussion of man's proper relationship with the physical creation as steward.
(Topics):book.

Walther, E. G. (1978). Stewardship and the food, energy, environment triangle. In M. E. Jegen & B. V. Manno (Eds.), *The Earth is the Lord's: essays on stewardship* (pp. 139-147). New York: Paulist Press.
(Topics):energy, stewardship, agriculture.

Wang, B. C. (1980). Demographic theories and policy positions on population and food. *Christian Scholar's Review*, 9(3), 241-255.
(Topics):population, agriculture.

Ward, B. (1972, July). Only One Earth. *Anticipation*, pp. 31+.
(Topics):popular.

Ward, B. (1973). *A new creation? Reflections on the environmental issue.* Vatican City: Pontifical Commission on Justice and Peace.
(Topics):popular book.

Ward, L. (1974). A six-letter obscenity. *Journal of the American Scientific Affiliation, 26*(1), 1-3.
(Topics):population.

Waskow, A. (1990). Ecology and social meaning: from compassion to jubilee. *Tikkun, 5*(2), 78-81.
(Topics):Judaism.

Watson, A. G. (1986). An International Storm Over Acid Rain. *Christian Century, 103*(16), 452-453.
(Topics):air pollution, acid rain.

Watty, W. W. (1981). Man and healing: a Biblical and theological view. *Point: Forum for Melanesian Affairs, 10*(2), 147-160.
There is a question whether this is the correct volume number. It could not be located in this study.
(Topics):popular.

Weaver, H. D., Jr. (1962). Introduction of symposium on the Christian's responsibility toward the increasing population. *Journal of the American Scientific Affiliation, 14*(1), 2.
(Topics):population.

Weaver, H. D., Jr. (1966). The moral issue of an expanding population. *Journal of the American Scientific Affiliation, 18*(4), 120.
(Topics):population.

Weber, L. (1987). Land use ethics: the social responsibility of ownership. In B. F. Evans & G. D. Cusack (Eds.), *Theology of the land* (pp. 13-39). Collegeville, MN: The Liturgical Press.
(Topics):land, ethics.

Weiss, D. W. (1983). The forces of nature, the forces of spirit: a perspective on Judaism. *Judaism*, 32(Fall), 477-487.
(Topics):Judaism.

Welbourn, F. B. (1975). Man's Dominion. *Theology*, 78(November), 561-568.
(Topics):theology.

Wendell, B. (1984). Two Economies. *Review and Expositor: A Baptist Theological Journal*, 81, 209-223.
The two "economies" are the human economy and the Kingdom of God.
(Topics):theology.

West, C. C. (1970). Theological guidelines for the future. *Theology Today*, 27(3), 277-291.
(Topics):theology.

West, C. C. (1975). Justice within the limits of the created world. *The Ecumenical Review*, 27(January), 57-64.
(Topics):popular.

West, C. C. (1981). God-women/man-creation: some comments on the ethics of the relationship. *The Ecumenical Review*, 33(January), 13-28.
(Topics):ethics.

Westermann, C. (1971). Creation and history in the Old Testament. In V. Vajta (Ed.), *The gospel and human destiny* (pp. 11-38). Minneapolis: Augsburg.
(Topics):theology.

Westermann, C. (1974). *Creation.* Philadelphia: Fortress Press.
Translation of an important German work. See especially pp. 51-55.
(Topics):theology.

Westermann, C. (1982). *Elements of Old Testament Theology*. Atlanta: John Knox Press.
(Topics):theology.

Westermann, C. (1984a). Biblical reflections on creator-creation. In B. W. Anderson (Ed.), *Creation in the Old Testament* (pp. 90-101). London: SPCK.
(Topics):theology.

Westermann, C. (1984b). *Genesis 1-11: a commentary* (Scullion, J. J., Trans.). Minneapolis: Augsburg Publishing House.
(Topics):theology.

Westermarck, E. (1939). *Christianity and morals*. New York: Macmillan.
See especially chapter 19.
(Topics):animal rights.

Westhelle, V. (1986). Labor: a suggestion for rethinking the way of the Christian. *Word & World: Theology for Christian Ministry*, 6(2), 194-206.
(Topics):ethics.

Westphal, M. (1982). Existentialism and environmental ethics. In E. R. Squiers (Ed.), *The environmental crisis: the ethical dilemma* (pp. 77-90). Mancelona, MI: Au Sable Institute of Environmental Studies.
(Topics):ethics.

Westwood, M. N. (1974). Limiting factors in world food supply and distribution. *Journal of the American Scientific Affiliation*, 26(3), 115-118.
Argues from a technology-will-solve-the-problem attitude and expresses little concern for the stewardly care of the environment.
(Topics):land, agriculture.

Wharton, R. (1987). Wilderness retreats: part one: wilderness manna. In F. W. Krueger (Ed.), *Christian ecology:*

building an environmental ethic for the twenty-first century (pp. 55). North Webster, Indiana: The North American Conference on Christianity and Ecology.
(Topics):wilderness.

White, H. (Ed.) (1964). *Christians in a technological era.* New York: Seabury Press.
(Topics):technology.

White, L., Jr. (1967). The Historical Roots of Our Ecologic Crisis. *Science*, 155, 1203-1207.
The paper that stirred up so much controversey in which White alleges that the Judeo-Christian tradition has contributed to the ecological crisis.
(Topics):popular.

White, L., Jr. (1973). Continuing the conversation. In I. Barbour (Ed.), *Western man and environmental ethics.* Reading, MA: Addison-Wesley.
Further discussion of the relationship between the Judeo-Christian faith and the deteriorating environment.
(Topics):popular.

Whitehouse, W. A. (1964). Towards a theology of nature. *Scottish Journal of Theology*, 17, 129-145.
(Topics):theology.

Wiant, H. V., Jr. (1980). Is clearcutting a responsible forestry practice? *Journal of the American Scientific Affiliation*, 32(4), 204-206.
(Topics):resources, trees.

Wicker, B. (1974). Ecology and the angels. *New Blackfriars*, 55, 4-15.
(Topics):popular.

Widman, R. (1970). When you've seen one beer can you've seen them all. *Eternity*, 21(5), 13-14+.
(Topics):popular, pollution.

Wilder, A. N. (1959). Eschatological imagery and earthly circumstance. *New Testament Studies*, 5(4), 229-245.
  Points out that our eschatological writings should deal with the redemption of both man and the creation.
  (Topics):theology.

Wilhelm, G. (1987). Youth's role in Christian Earth stewardship. In F. W. Krueger (Ed.), *Christian ecology: building an environmental ethic for the twenty-first century* (pp. 104-105). North Webster, Indiana: The North American Conference on Christianity and Ecology.
  (Topics):popular.

Wilkinson, L. (1975). Christian ecology of death: biblical imagery and the ecologic crisis. *Christian Scholar's Review*, 5(4), 319-338.
  (Topics):popular.

Wilkinson, L. (Ed.) (1980a). *Earthkeeping: Christian stewardship of natural resources*. Grand Rapids: Eerdmans Publishing Co.
  See especially chapters 6 - 9.
  (Topics):ethics, history, theology, economics.

Wilkinson, L. (1980b). Global housekeeping: lords or servants: we reflect the duality of the incarnation in our relationship to the Earth. *Christianity Today*, 24(12), 752-756.
  (Topics):popular.

Wilkinson, L. (1981). Cosmic Christology and the Christian's role in creation. *Christian Scholar's Review*, 11(1), 18-41.
  (Topics):theology, popular.

Wilkinson, L. (1982). Redeemers of the Earth. In E. R. Squiers (Ed.), *The environmental crisis: the ethical dilemma* (pp. 39-56). Mancelona, MI: Au Sable Institute of Environmental Studies.
  (Topics):ethics, popular.

Wilkinson, L. (1987a). New age, new consciousness, and the new creation. In W. Granberg-Michaelson (Ed.), *Tending the garden: essays on the gospel and the Earth* (pp. 6-29). Grand Rapids: Eerdmans.
    A response to Cumbey 1983 and Hunt 1983 who argue that the Christian faith is not concerned with creation care. Those Christians who take a pro-environment position are labeled as part of the New Age.
    (Topics):new age.

Wilkinson, L. (1987b). A response to stewardship of the environment. In K. S. Kantzer (Ed.), *Applying the scriptures: papers from ICBI summit III* (pp. 497-502). Grand Rapids, MI: Academie Books, Zondervan Publishing House.
    (Topics):theology.

Wilkinson, L. (1990). A theology of the beasts. *Christianity Today*, 34(9), 21.
    (Topics):animal rights.

Wilkinson, L. (1991). Christ as creator and redeemer: scriptural teachings on cosmic Christology. In C. B. DeWitt (Ed.), *The environment and the Christian: what does the New Testament say about the environment?* Grand Rapids, Michigan: Baker Book House.
    (Topics):theology.

Williams, D. L. (1970). Our environment: a challenge to reason. *Church and Society*, 60(January/February), 19-26.
    (Topics):popular.

Williams, G. H. (1962). *Wilderness and paradise in Christian thought: the Biblical experience of the desert in the history of Christianity & the paradise theme in the theological idea of the university*. New York: Harper & Brothers Publishers.
    (Topics):theology, wilderness.

Williams, G. H. (1970). The background in scripture and tradition for reassessment of a Christian role in wildlife

conservation and the stewardship of natural resources. *Colloquy*, 3(4), 12-15.
    (Topics):biodiversity, resources.

Williams, G. H. (1971). Christian attitudes toward nature. *Christian Scholar's Review*, 2(1), 3-35.
    Response to the charges of White 1967 and others.
    (Topics):popular, theology.

Williams, G. H. (1972). Christian attitudes toward nature. *Christian Scholar's Review*, 2(2), 112-126.
    Part two of Williams 1971.
    (Topics):popular, theology.

Williams, J. (1989-90). Recycling in Houston: the churches lead. *The Egg: A Journal of Eco-Justice*, 9(4), 9-10.
    (Topics):resources, recycling.

Willis, D. (1983). Priests of creation: who will represent God to the trees? *The Other Side*, 142(July), 24-25.
    (Topics):popular.

Willis, D. L. (1980). Nukes or no nukes? Absolute thinking in a relative world. *Journal of the American Scientific Affiliation*, 32(June), 102-108.
    (Topics):nuclear.

Willis, E. D. (1985). Proclaiming liberation for the Earth's sake. In D. Hessel (Ed.), *For Creation's sake: preaching, ecology and justice* (pp. 55-70). Philadelphia: Geneva Press.
    (Topics):theology.

Wingren, G. (1961). *Creation and Law*. Philadelphia: Muhlenberg Press.
    Portrays a utilitarian view of the man/nature relationship.
    (Topics):theology.

Wingren, G. (1984). The doctrine of creation: not an appendix but the first article. *Word & World*, 4(4), 353-371.

(Topics):theology.

Wink, W. (1978). The "elements of the universe" in biblical and scientific perspective. *Zygon*, 13(3), 225-248.
(Topics):theology.

Winn, A. C. (1985). Holding the Earth in trust. In N. C. Murphy (Ed.), *Teaching and preaching stewardship: an anthology* (pp. 210-216). New York: Commission on Stewardship, National Council of the Churches of Christ in the U.S.A.
(Topics):popular, stewardship.

Winter, G. (1981). *Liberating Creation: foundations of religious social ethics.* New York: Crossroad.
(Topics):popular.

Wise, D. S. (1989). Environmental stewardship literature and the New Testament: a review. In C. B. DeWitt (Ed.), *The environment and the Christian: what does the New Testament say about the environment?* Grand Rapids, Michigan: Baker Book House.
(Topics):bibliography.

Wolters, A. (1984). Nature and grace in the interpretation of proverbs 31:10-31. *Calvin Theological Journal*, 19(November), 153-166.
(Topics):theology.

Wolters, A. (1985). *Creation regained: Biblical basics for a reformational worldview.* Grand Rapids, MI: Eerdmans Publishing Company.
Emphasis is on world views. The physical creation is a minor subject.
(Topics):popular book.

Wolterstorff, N. (1983). *Until justice and peace embrace.* Grand Rapids, MI: Eerdmans.

Although the environment is not the central thrust of this excellent work, it is clearly linked by the author to the global justice and peace issue.
(Topics):popular book.

Wood, A. (1964). Intimacy of Jesus with nature. *London Quarterly and Holborn Review*, 189(January), 44-51.
(Topics):theology.

Wood, B. (1986). *Our world, God's world: reflections for Advent and the Christmas season on the environment.* Cincinnati, OH: Forward Movement Publications.
(Topics):popular book.

World Council of Churches (1971). Global environment, responsible choice and social justice. *The Ecumenical Review*, 23(October), 438-442.
(Topics):popular.

World Council of Churches (1974a). *The churches in international affairs: reports 1970-1973.* World Council of Churches.
Commission of the Churches on International Affairs.
(Topics):popular.

World Council of Churches (1974b). Threats to Survival. *Study Encounter*, 10(4).
Unable to determine page numbers; the work was not seen in this study.
(Topics):popular.

World Council of Churches (1984). Theological aspects. *The Ecumenical Review*, 36, 33-42.
(Topics):popular.

World Council of Churches (1989). *Peace with justice.* Final document of the European Assembly, Basel, Switzerland, 15-21 May 1989. Available at the Conference of European Churches Office, 150 route de Ferney, 1211 Geneva 2.

(Topics):popular, land.

World Council of Churches (1990). *Now is the time: final document & other texts*. The final document of the World Convocation on Justice, Peace and the Integrity of Creation held in Seoul, Korea 5-12 March 1990. Copies may be ordered from JPIC-Office in the WCC, P.O. Box 2100, CH-1211 Geneva 2. Cost $3.00 (U.S.).
(Topics):popular.

Worster, D. (1977). *Nature's economy: the roots of ecology*. San Francisco: Sierra Club Books.
(Topics):history.

Wright, C. J. H. (1990). *God's people in God's land: family, land, and property in the Old Testament*. Grand Rapids: Wm. B. Eerdmans Pub. Co.
(Topics):land, theology.

Wright, R. T. (1970). Responsibility for the ecological crisis. *BioScience*, 20(15), 851-853.
Reprinted in *Christian Scholar's Review*, 1(1):35-40.
(Topics):popular.

Wright, R. T. (1989). *Biology through the eyes of faith*. San Francisco: Harper & Row, Publishers.
One of several books focusing on academic disciplines that are being published by the Christian College Coalition. They are designed to be used as supplementary reading to allow integration of faith in conjunction with a secular text book. See especially chapters 1, 2, 8, 9, 12, and 13.
(Topics):popular.

Wrightsman, B. (1970). Man: manager or manipulator of the Earth. *Dialog*, 9(Summer), 200-214.
(Topics):history.

Wyatt, S. P. (1983). *Jesus Christ and Creation in the theology of John Calvin*. Th.D., Toronto School of Theology.
(Topics):theology.

Wylie, H. G. (1988). The farmer's dilemmas. *Touchstone*, 6(May), 41-43.
(Topics):land.

Yamauchi, E. M. (1974). Exaggerated, Radical and Unrealistic. *Journal of the American Scientific Affiliation*, 26(1), 21.
(Topics):population.

Yamauchi, E. M. (1980). Ancient ecologies and the biblical perspective. *Journal of the American Scientific Affiliation*, 32(4), 193-203.
(Topics):history.

Yancey, P. (1991). A voice crying in the rainforest. *Christianity Today*, 35(8), 26-28.
One of the world's leading botanists, Ghillean Prance, believes "God so loved the world" carries with it a holy obligation to preserve what God has made.
(Topics):biodiversity, popular.

Yandell, K. (1982). Fundamentals of environmental ethics: east and west. In E. R. Squiers (Ed.), *The environmental crisis: the ethical dilemma* (pp. 91-108). Mancelona, MI: Au Sable Institute of Environmental Studies.
(Topics):ethics.

Yoder, J. H. (1979). Response to Scott Paradise paper: vision of a good society. *Anglican Theological Review*, 61(1), 118-126.
(Topics):energy.

Yoder, J. L. (1978). *The ethics of eco-justice: a Christian response to the ecological crisis*. Ph.D., Duke University.
(Topics):theology, ethics, ecojustice.

Yoder, R. A. (1983). *Seeking first the kingdom, called to faithful stewardship*. Scottdale, Pennsylvania: Herald Press.

(Topics):popular book, stewardship.

Young, N. (1976). *Creator, creation and faith.* Philadelphia: Westminster Press.
(Topics):theology.

Young, R. V. (1974). Christianity and ecology. *National Review*, 26, 1454-1458+.
(Topics):popular.

Young, R. V., Jr. (1979). A conservative view of environmental affairs. *Environmental Ethics*, 1(3), 241-254.
(Topics):ethics.

Yutzy, V. (1991). Cooperating with Creation. *Earthkeeping: a quarterly on faith and agriculture*, 6(3), 10-12
(Topics):land, agriculture.

Zaidi, I. H. (1981). On the ethics of man's interaction with the environment: an Islamic approach. *Environmental Ethics*, 3(1), 33-47.
(Topics):Islam.

Zelesnik, P. (1987). Eco-feminism and the church: part two: feminism as human liberation. In F. W. Krueger (Ed.), *Christian ecology: building an environmental ethic for the twenty-first century* (pp. 99). North Webster, Indiana: The North American Conference on Christianity and Ecology.
(Topics):feminism.

Zerbe, G. (1991). The meaning of the Kingdom of God for the stewardship of Creation. In C. B. DeWitt (Ed.), *The environment and the Christian: what does the New Testament say about the environment?* Grand Rapids, Michigan: Baker Book House.
(Topics):theology.

Zinkand, D. (1986). Colleagues, not competitors in agriculture. *Earthkeeping: A Quarterly on Faith and Agriculture*, 2(4), 15.

(Topics):land, agriculture.

Zizioulas, J. (1989). Preserving God's creation: three lectures on theology and ecology; pt. 1 and pt. 2. *King's Theological Review*, 12(Spring), 1-5; 12(Autumn), 41-45.
(Topics):theology, history.

Zizioulas, J. (1990). Preserving God's creation: three lectures on theology and ecology; pt. 3. *King's Theological Review*, 13(Spring), 1-5.
(Topics):theology, history.

Zohary, M. (1982). *Plants of the Bible: a complete handbook.* New York: Cambridge University Press.
(Topics):scriptural plants and animals.

Zygon (thematic issue) (1970). Conference on ethics and ecology of the Institute on Religion in the Age of Science. *Zygon*, 5(4), 270-351.
(Topics):popular, ethics.

Zylstra, U. (1978, July). Tending God's garden: the Christian in agriculture today. *Reformed Journal*, pp. 9-11.
(Topics):agriculture, land.

Zylstra, U. (1982). Ecological aspects of good production: biblical directives for agriculture. In E. R. Squiers (Ed.), *The environmental crisis: the ethical dilemma* (pp. 135-150). Mancelona, MI: Au Sable Institute of Environmental Studies.
(Topics):land.